Modern Quantum Chemistry

Modern Quantum Chemistry

Declan Hicks

NY RESEARCH PRESS

New York

Published by NY Research Press
118-35 Queens Blvd., Suite 400,
Forest Hills, NY 11375, USA
www.nyresearchpress.com

Modern Quantum Chemistry
Declan Hicks

International Standard Book Number: 978-1-63238-787-5 (Hardback)

Cataloging-in-Publication Data

Modern quantum chemistry / Declan Hicks.
    p. cm.
Includes bibliographical references and index.
ISBN 978-1-63238-787-5
1. Quantum chemistry. 2. Quantum theory. 3. Chemistry,
Physical and theoretical. I. Hicks, Declan.
QD462 .M63 2020
541.28--dc23

Table of Contents

It is with great pleasure that I present this book. It has been carefully written after numerous discussions with my peers and other practitioners of the field. I would like to take this opportunity to thank my family and friends who have been extremely supporting at every step in my life.

The branch of chemistry which focuses on the application of quantum mechanics in physical models is referred to as quantum chemistry. It deals with the study of ground state of individual molecules and atoms as well as the transition and excited states that occur during chemical reactions. The two major branches of quantum chemistry are experimental quantum chemistry and theoretical quantum chemistry. Experimental quantum chemistry seeks to obtain information regarding the quantization of energy on a molecular scale. It relies extensively on spectroscopy. Nuclear magnetic resonance spectroscopy, infrared spectroscopy and scanning probe microscopy are a few of the common methods used in this area. The predictions of quantum theory are calculated within the field of theoretical quantum chemistry. This book aims to shed light on some of the unexplored aspects of quantum chemistry. While understanding the long-term perspectives of the topics, it makes an effort in highlighting their impact as a modern tool for the growth of the discipline. Those in search of information to further their knowledge will be greatly assisted by this book

The chapters below are organized to facilitate a comprehensive understanding of the subject:

Chapter – What is Quantum Chemistry?

Quantum chemistry is the branch of chemistry that deals with the application of quantum mechanics to the study of molecules. Hund's rule and Aufbau principle fall under its domain. This is an introductory chapter which will briefly introduce about quantum chemistry.

Chapter – Many-electron Atoms

Many-electron atom refers to an atom which has more than 1 electron. It includes the aspects of spin quantum number, azimuthal quantum number, magnetic quantum number, angular momentum of electrons, etc. The topics elaborated in this chapter will help in gaining a better perspective about these aspects related to many-electron atoms.

Chapter – Lone Pair of Electrons

Lone pair of electrons is the pair of valence electrons that are not shared with another atom in a covalent bond. A single lone pair can be found in the nitrogen group, two lone pairs can be found in the chalcogen group and the halogens can carry three lone pairs. This chapter has been carefully written to provide an easy understanding of the related aspects of lone pair of electrons.

Chapter – Types of Bonds

Bond is referred to as the attraction between atoms, ions and molecules to form a chemical compound. It can be categorized into pi bond, sigma bond, single bond, double bond and triple bond. This chapter delves into these types of bonds to provide an in-depth understanding of the subject.

Chapter – Theories in Quantum Chemistry

Some of the theories used in quantum chemistry are valence bond theory, molecular orbit theory, VSEPR theory, density functional theory, time-dependent density functional theory, etc. This chapter closely examines these theories of quantum chemistry to provide an extensive understanding of the subject.

Chapter – Qualitative Theories of Chemical Bonding

Qualitative theory of chemical bonding includes atomic orbitals, molecular orbitals, electronic configuration, Huckel approximation and Huckel molecular orbital theory. The topics elaborated in this chapter will help in gaining a better perspective about these qualitative theories of chemical bonding.

Chapter – Diverse Aspects of Quantum Chemistry

Some of the important concepts of quantum chemistry include Born–Oppenheimer approximation, Zeeman effect, Brillouin's theorem, Koopmans' theorem, Schrödinger equation, etc. This chapter closely examines these key concepts of quantum chemistry to provide an extensive understanding of the subject.

Chapter – Computational Quantum Chemistry

Computational quantum chemistry is one of the fields of chemistry that makes use of computer programs and approximations for solving chemical problems. It calculates electronic charge density, dipoles and multiple moments, absolute and relative energies, vibrational frequencies of molecules and solids. This chapter discusses this field of computational quantum chemistry in detail.

Declan Hicks

1

What is Quantum Chemistry?

Quantum chemistry is the branch of chemistry that deals with the application of quantum mechanics to the study of molecules. Hund's rule and Aufbau principle fall under its domain. This is an introductory chapter which will briefly introduce about quantum chemistry.

Quantum chemistry is the application of quantum mechanical principles and equations to the study of molecules. In order to understand matter at its most fundamental level, we must use quantum mechanical models and methods. There are two aspects of quantum mechanics that make it different from previous models of matter. The first is the concept of wave-particle duality; that is, the notion that we need to think of very small objects (such as electrons) as having characteristics of both particles and waves. Second, quantum mechanical models correctly predict that the energy of atoms and molecules is always quantized, meaning that they may have only specific amounts of energy. Quantum chemical theories allow us to explain the structure of the periodic table, and quantum chemical calculations allow us to accurately predict the structures of molecules and the spectroscopic behavior of atoms and molecules.

Quantum mechanical ideas began with studies of the physics of light. By the late nineteenth century, virtually all scientists believed that light behaved as a wave. Although some earlier scientists, such as Isaac Newton in the seventeenth century, had thought of light as consisting of particles, the early nineteenth-century experiments of Thomas Young and Augustin Fresnel demonstrated that light has wavelike properties. In these experiments, light was passed through a pair of slits in a screen, and produced alternating light and dark regions (interference patterns) on a second screen. This phenomenon, known as diffraction, cannot be explained using a particle model for light. In the late nineteenth century, James Clerk Maxwell derived a set of equations based on the wave model for light, which beautifully explained most experimental results.

Despite this apparent certainty that light was a wave, Max Planck and Albert Einstein, at the beginning of the twentieth century, showed that some experiments required the use of a particle model for light, rather than a wave model. Since both models were necessary for an accurate description of all of the properties of light, scientists today use mathematical equations appropriate to both waves and particles in describing the properties of light.

Waves and particles are fundamentally different: a particle exists at a particular point in space, whereas a wave continues on for (sometimes) a great distance. It defies intuition to think that both of these models might describe the same thing. Nevertheless, an accurate description of light requires the use of both wave and particle ideas.

The Wave Nature of Matter

The success of wave-particle duality in describing the properties of light paved the way for using that same idea in describing matter. Experiments in the early twentieth century showed that the energy in atoms is quantized, a given atom can have only specific amounts of energy. For hydrogen, the simplest of the atoms, an accurate formula for the possible energies had been experimentally determined but was unexplainable using any particle model for the atom. The best picture that the particle model could give, consistent with experiments on atoms, put the electron in a sort of "orbit" around the nucleus. Unfortunately, the particle model predicts that the electron should collide with the nucleus, releasing energy in the process. Obviously there was a need for a different model for the electron.

One example of quantization would be to think of your distance from the ground when standing on a ladder, your distance from the ground can only change one rung at a time.

In 1924 Louis de Broglie presented a theory for the hydrogen atom that modeled the electron as a wave. Calculations made for this model give the quantization of energy that is experimentally observed in this atom. De Broglie also postulated a general formula for obtaining the wavelength of a moving object. His formula, which is analogous to that used for light, states that the wavelength of a moving object is inversely proportional to its momentum (mass times velocity). When one uses de Broglie's formula to determine wavelengths of macroscopic objects, one discovers that the wavelengths of even the smallest objects visible to the naked eye are too small for the wavelike characteristics of these objects to be significant in any real situation. For the electron, however, the wavelength is large enough to be measurable. Diffraction experiments have been performed using electrons, demonstrating conclusively that they have wave properties.

It is contrary to our intuition that electrons might behave as waves. The repercussions of this notion is that the electron does not have a definite size, but is spread out over a region in space. We are more comfortable with the thought of the electron being a microscopic particle, moving around in an orbit near the nucleus of an atom. As with light, however, we do not abandon the particle model for electrons; rather, we employ mathematical equations arising from both particle and wave models. For quantum chemical calculations, the wave model turns out to be more useful.

The Heisenberg Uncertainty Principle

One consequence of the wave nature of matter is that the position and momentum of small objects are not well known, as they would be for a particle model. In some circumstances, a wave may be confined to a very narrow region in space; however, there is still some uncertainty as to its position. Additionally, the value of the momentum of a quantum object is often not known precisely. In 1927 Werner Heisenberg showed that the product of the uncertainty in position and the uncertainty in momentum is greater than or equal to a certain constant (Planck's constant divided by 4π). This constant is very small; accordingly, quantum mechanical uncertainty in position and momentum of objects that are large enough to see is not noticed experimentally. For electrons, however, quantum mechanical uncertainties in position and momentum are important considerations in interpreting both theoretical models and experimental results. The relationship between the uncertainties in position and momentum is known as the Heisenberg Uncertainty Principle. It tells us that the more we know about the position of a small object, such as an electron, the less we know about its momentum (and vice versa).

Calculating the Wavefunction

For a scientist, knowing that matter behaves as a wave is useful only if one knows something about that wave. The wavefunction is a mathematical function describing the wave. For example, $y(x) = A \sin(kx)$ might be the wave-function for a one-dimensional wave, which exists along the x-axis. Matter waves are three-dimensional; the relevant wavefunction depends on the x, y, and z coordinates of the system being studied (and sometimes on time as well). We conventionally label the wavefunction for a three-dimensional object as $\psi(x, y, z)$. In 1926 Erwin Schrödinger introduced a mathematical equation whereby, if one knows the potential energy acting on an object, one can calculate the wavefunction for that object. Heisenberg had already introduced a mathematical formalism for performing quantum mechanics calculations, without explicitly including the concept of waves. It was later shown that, although the approaches of Schrödinger and Heisenberg looked very different, they made exactly the same predictions. In practice, the Schrödinger formalism is more useful for explaining the problem being studied, and the Heisenberg methodology allows for more straightforward computation. Accordingly, a mixture of the two approaches is typically used in modern quantum chemistry. Once we know the wavefunction of the atom or molecule under study, we can calculate the properties of that atom or molecule.

Quantum Mechanics of Atoms

An exact solution for Schrödinger's wave equation can be obtained for the hydrogen atom; however, for larger atoms and molecules (which contain more than one electron), Schrödinger's equation can be solved only approximately. Although this may sound so restrictive as to make the equation useless, there are well-established approaches that allow for practical and accurate calculations on atoms and molecules. This is done by making some assumptions about larger systems based upon the hydrogen atom, as explained below.

When the Schrödinger equation is solved for the hydrogen atom, the resulting wavefunctions for the various possible energies that the atom can have are used to determine atomic orbitals. An orbital is a region in space where an electron is most likely to be found. For example, the lowest-energy wavefunction for a hydrogen atom is the so-called 1s orbital, which is a spherical region in space surrounding the nucleus. For some higher-energy states, the orbitals are not necessarily spherical in shape.

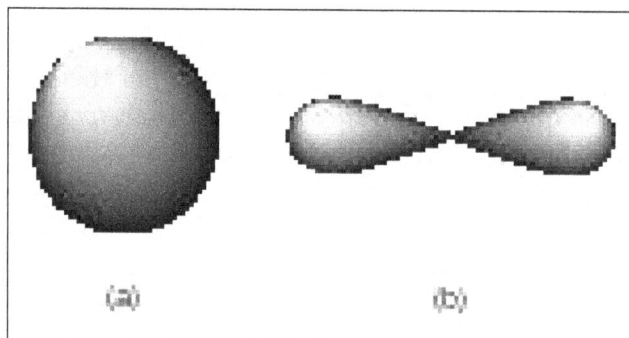

For atoms larger than hydrogen, one assumes that the orbitals occupied by the electrons have the same shape as the hydrogen orbitals but are differing in size and energy. The energies corresponding to these orbitals may be found by solving an approximate version of Schrödinger's equation. These atomic orbitals, in turn, may be used as the building blocks to the electronic behavior in molecules, As it happens, two electrons may share an atomic orbital; we say that these electrons are paired. Chemists have developed a system of rules for determining which orbitals are occupied in which atoms; calculations can then be done to determine the energies of the electrons in the atoms.

Quantum Mechanics of Molecules

Molecules are held together by covalent bonds. The simplest definition of a covalent bond is a shared pair of electrons. There are two basic approaches to modeling covalent bonds in molecules: the valence bond model and the molecular orbital model. In the valence bond model, we think of atomic orbitals on each of two atoms combining to form a bond orbital, with one electron from each atom occupying this orbital. Both the bond orbital and the electron pair now "belong" to both of the atoms. This sharing of electrons brings about a lowering in the energy, which makes the formation of molecules from atoms an energetically favorable process. The valence bond model gives the simplest quantum mechanical picture of chemical bonding, but it is not the best method for accurate calculations on molecules containing more than two atoms.

Molecular orbital theory differs from valence bond theory in that it does not require the electrons involved in a bond to be localized between two of the atoms in a molecule. Instead, the electron occupies a molecular orbital, which may be spread out over the entire molecule. As in the valence bond approach, the molecular orbital is formed by adding up contributions from the atomic orbitals on the atoms that make up the molecule. This approach, which does not explicitly model bonds as existing between two atoms, is somewhat less appealing to the intuition than the valence bond approach. However, molecular orbital calculations typically yield better predictions of molecular structure and properties than valence bond methods. Accordingly, most commercially available quantum chemistry software packages rely on molecular orbital methods to perform calculations.

Schrödinger's Wave Equation

Schrödinger's Wave equation may be written (in abbreviated form) as:

$$\hat{E}_K \psi\left(x,\ y,\ z\right) + \hat{E}_P \psi\left(x,\ y,\ z\right) = \hat{E}\psi\left(x,\ y,\ z\right)$$

The first term, $\hat{E}_K \psi\left(x,\ y,\ z\right)$, represents the kinetic energy of the system being studied. The second term, $\hat{E}_P \psi\left(x,\ y,\ z\right)$, represents the potential energy of the system. E and $\psi\left(x,\ y,\ z\right)$ are the total energy of the system and wavefunction describing the system, respectively. Once the wavefunction is determined, virtually any property of the molecule may be calculated.

A lot of the modern research in quantum chemistry is focused on improving the valence bond and molecular orbital methods for calculating molecular properties. Different underlying approximations and different orbital functions are tried, and the results are compared with previous calculations and with experimental data to determine which methods give the best results. It is often the case that the best choice of quantum chemical method depends on the particular molecule or molecular property being studied.

Applications of Quantum Chemistry to Chemical Structure and Reactivity

The explosion in popularity of density functional theory (DFT) has created an opportunity for critical evaluation of the performance of DFT methods versus accurate quantum and experimental data. Quantities such as HOMO-LUMO gaps and reaction barriers (transition states) are among the well-known cases where DFT methods often come up short. Careful validation studies will establish what functionals yield the most accurate results and reveal systematic deficiencies. Such studies open the door to the development of improved functionals.

Quantum chemistry continues to show its value in supporting and interpreting experimental spectroscopic data. A close cooperation between theoretical calculations and experiments has yielded a number of opportunities for quantum chemistry calculations to clearly identify species found in spectra and to suggest new avenues for experimental study.

As the number of atoms in a molecule increases, calculations become correspondingly expensive. One means of addressing this problem is through the development and parameterization of semi-empirical methods. A tight binding code has been developed in our division and in collaboration with other researchers. Tight binding methods have proven to yield reliable results on larger molecular systems with suitable parameterization. We are studying ways to automate parameter generation to facilitate further studies. Applications of the tight binding methodology include (but are not limited to) biomolecules, transition metal surfaces, and nanoparticles.

Chemical Reactivity Theory provides a framework for the calculation of a number of indices which characterize chemical reactivity (e.g., Fukui function, electrophilicity). These indices can be correlated to reactive sites with a molecule and can be used to predict rates of chemical reactions. We are applying these methodologies in such diverse areas as the calculation of global warming potentials and of rates of chemical reactions on nanoparticle surfaces.

Simulation of reactive processes on metallic surfaces and on the surfaces of metallic nanoparticles in an aqueous environment and with potential control (electrochemistry) is an extremely challenging problem at the forefront of computational chemistry. We are beginning to move into this new field in collaboration with an experimental electrochemistry group. Among the many application areas are clear energy generation and storage.

Ab Initio Quantum Chemistry Methods

Ab initio quantum chemistry methods are computational chemistry methods based on quantum chemistry. The term ab initio was first used in quantum chemistry by Robert Parr and coworkers, including David Craig in a semiempirical study on the excited states of benzene. The background is described by Parr. Ab initio means "from first principles" or "from the beginning", implying that the only inputs into an ab initio calculation are physical constants. Ab initio quantum chemistry methods attempt to solve the electronic Schrödinger equation given the positions of the nuclei and the number of electrons in order to yield useful information such as electron densities, energies and other properties of the system. The ability to run these calculations has enabled theoretical chemists to solve a range of problems and their importance is highlighted by the awarding of the Nobel prize to John Pople and Walter Kohn.

Accuracy and Scaling

Ab initio electronic structure methods aim to calculate the many electron function which is the solution of the non-relativistic solution of the electronic Schrödinger equation (in the Born-Oppenheimer approximation). The many electron function is generally a linear combination of many simpler electron functions with the dominant function being the Hartree-Fock function. Each of these simple functions are then approximated using only one-electron functions. The one-electron functions are then expanded as a linear combination of a finite set of basis functions. This approach has the advantage that it can be made to converge to the exact solution, when the basis set tends toward the limit of a complete set and where all possible configurations are included (called "Full CI"). However, this convergence to the limit is computationally very demanding and most calculations are far from the limit. Nevertheless important conclusions have been made from these more limited classifications.

One needs to consider the computational cost of ab initio methods when determining whether they are appropriate for the problem at hand. When compared to much less accurate approaches, such as molecular mechanics, ab initio methods often take larger amounts of computer time, memory, and disk space, though, with modern advances in computer science and technology such considerations are becoming less of an issue. The Hartree-Fock (HF) method scales nominally as N^4 (N being a relative measure of the system size, not the number of basis functions) - e.g., if one doubles the number of electrons and the number of basis functions (double the system size), the calculation will take 16 (2^4) times as long per iteration. However, in practice it can scale closer to N^3 as the program can identify zero and extremely small integrals and neglect them. Correlated calculations scale less favorably, though their accuracy is usually greater, which is the trade off one needs to consider. One popular method is Møller-Plesset perturbation theory (MP). To second order (MP2), MP scales as N^4. To third order (MP3) MP scales as N^6. To fourth order (MP4) MP scales as N^7. Another method, coupled cluster with singles and doubles (CCSD), scales as N^6 and extensions, CCSD(T) and CR-CC(2,3), scale as N^6 with one noniterative step which scales as N^7. Density functional theory (DFT) methods using functionals which include Hartree-Fock exchange scale in a similar manner to Hartree-Fock but with a larger proportionality term and are thus more expensive than an equivalent Hartree-Fock calculation. DFT methods that do not include Hartree-Fock exchange can scale better than Hartree-Fock.

Linear Scaling Approaches

The problem of computational expense can be alleviated through simplification schemes. In the density fitting scheme, the four-index integrals used to describe the interaction between electron pairs are reduced to simpler two- or three-index integrals, by treating the charge densities they contain in a simplified way. This reduces the scaling with respect to basis set size. Methods employing this scheme are denoted by the prefix "df-", for example the density fitting MP2 is df-MP2 (many authors use lower-case to prevent confusion with DFT). In the local approximation, the molecular orbitals are first localized by a unitary rotation in the orbital space (which leaves the reference wave function invariant, i.e., is not an approximation) and subsequently interactions of distant pairs of localized orbitals are neglected in the correlation calculation. This sharply reduces the scaling with molecular size, a major problem in the treatment of biologically-sized molecules. Methods employing this scheme are denoted by the prefix "L", e.g. LMP2. Both schemes can be employed together, as in the df-LMP2 and df-LCCSD(T0) methods. In fact, df-LMP2 calculations are faster than df-Hartree-Fock calculations and thus are feasible in nearly all situations in which also DFT is.

Classes of Methods

The most popular classes of ab initio electronic structure methods:

Hartree-Fock Methods

- Hartree-Fock (HF).

- Restricted open-shell Hartree-Fock (ROHF).

- Unrestricted Hartree-Fock (UHF).

Post-Hartree-Fock Methods

- Møller-Plesset perturbation theory (MPn).

- Configuration interaction (CI).

- Coupled cluster (CC).

- Quadratic configuration interaction (QCI).

- Quantum chemistry composite methods.

Multi-Reference Methods

- Multi-configurational self-consistent field (MCSCF including CASSCF and RASSCF).

- Multi-reference configuration interaction (MRCI).

- n-electron valence state perturbation theory (NEVPT).

- Complete active space perturbation theory (CASPTn).

- State universal multi-reference coupled-cluster theory (SUMR-CC).

Methods in Detail

Hartree-Fock and Post-Hartree-Fock Methods

The simplest type of ab initio electronic structure calculation is the Hartree-Fock (HF) scheme, in which the instantaneous Coulombic electron-electron repulsion is not specifically taken into account. Only its average effect (mean field) is included in the calculation. This is a variational procedure; therefore, the obtained approximate energies, expressed in terms of the system's wave function, are always equal to or greater than the exact energy, and tend to a limiting value called the Hartree-Fock limit as the size of the basis is increased. Many types of calculations begin with a Hartree-Fock calculation and subsequently correct for electron-electron repulsion, referred to also as electronic correlation. Møller-Plesset perturbation theory (MPn) and coupled cluster theory (CC) are examples of these post-Hartree-Fock methods. In some cases, particularly for bond breaking processes, the Hartree-Fock method is inadequate and this single-determinant reference function is not a good basis for post-Hartree-Fock methods. It is then necessary to start with a wave function that includes more than one determinant such as multi-configurational self-consistent field (MCSCF) and methods have been developed that use these multi-determinant references for improvements. However, if one uses coupled cluster methods such as CCSDT, CCSDt, CR-CC(2,3), or CC(t;3) then single-bond breaking using the single determinant HF reference is feasible. For an accurate description of double bond breaking, methods such as CCSDTQ, CCSDTq, CCSDtq, CR-CC(2,4), or CC(tq;3,4) also make use of the single determinant HF reference, and do not require one to use multi-reference methods.

Example: Is the bonding situation in disilyne Si_2H_2 the same as in acetylene (C_2H_2)?

A series of ab initio studies of Si_2H_2 is an example of how ab initio computational chemistry can predict new structures that are subsequently confirmed by experiment. They go back over 20 years, and most of the main conclusions were reached by 1995. The methods used were mostly post-Hartree-Fock, particularly configuration interaction (CI) and coupled cluster (CC). Initially the question was whether disilyne, Si_2H_2 had the same structure as ethyne (acetylene), C_2H_2. In early studies, by Binkley and Lischka and Kohler, it became clear that linear Si_2H_2 was a transition structure between two equivalent trans-bent structures and that the ground state was predicted to be a four-membered ring bent in a 'butterfly' structure with hydrogen atoms bridged between the two silicon atoms. Interest then moved to look at whether structures equivalent to vinylidene ($Si=SiH_2$) existed. This structure is predicted to be a local minimum, i. e. an isomer of Si_2H_2, lying higher in energy than the ground state but below the energy of the trans-bent isomer. Then a new isomer with an unusual structure was predicted by Brenda Colegrove in Henry F. Schaefer, III's group. It requires post-Hartree-Fock methods to obtain a local minimum for this structure. It does not exist on the Hartree-Fock energy hypersurface. The new isomer is a planar structure with one bridging hydrogen atom and one terminal hydrogen atom, cis to the bridging atom. Its energy is above the ground state but below that of the other isomers. Similar results were later obtained for Ge_2H_2. Al_2H_2 and Ga_2H_2 have exactly the same isomers, in spite of having two electrons less than the Group 14 molecules. The only difference is that the four-membered ring ground state is planar and not bent. The cis-mono-bridged and vinylidene-like isomers are present. Experimental work on these molecules is not easy, but matrix isolation spectroscopy of the products of the reaction of hydrogen atoms and silicon and aluminium surfaces has found the ground state ring structures and the cis-mono-bridged structures for Si_2H_2 and Al_2H_2. Theoretical predictions of the vibrational frequencies were crucial in understanding the experimental observations of the spectra of a mixture of compounds. This may appear to be an obscure area of chemistry, but the differences between carbon and silicon chemistry is always a lively question, as are the differences between group 13 and group 14 (mainly the B and C differences).

Valence Bond Methods

Valence bond (VB) methods are generally ab initio although some semi-empirical versions have been proposed. Current VB approaches are:-

- Generalized valence bond (GVB).

- Modern valence bond theory (MVBT).

Quantum Monte Carlo Methods

A method that avoids making the variational overestimation of HF in the first place is Quantum Monte Carlo (QMC), in its variational, diffusion, and Green's function forms. These methods work with an explicitly correlated wave function and evaluate integrals numerically using a Monte Carlo integration. Such calculations can be very time-consuming. The accuracy of QMC depends strongly on the initial guess of many-body wave-functions and the form of the many-body wave-function. One simple choice is Slater-Jastrow wave-function in which the local correlations are treated with the Jastrow factor.

Hund's Rule

Hund's rule states that:

- Every orbital in a sublevel is singly occupied before any orbital is doubly occupied.

- All of the electrons in singly occupied orbitals have the same spin (to maximize total spin).

When assigning electrons to orbitals, an electron first seeks to fill all the orbitals with similar energy (also referred to as degenerate orbitals) before pairing with another electron in a half-filled orbital. Atoms at ground states tend to have as many unpaired electrons as possible. In visualizing this process, consider how electrons exhibit the same behavior as the same poles on a magnet would if they came into contact; as the negatively charged electrons fill orbitals, they first try to get as far as possible from each other before having to pair up.

Example: Nitrogen Atoms.

Consider the correct electron configuration of the nitrogen (Z = 7) atom: $1s^2 2s^2 2p^3$:

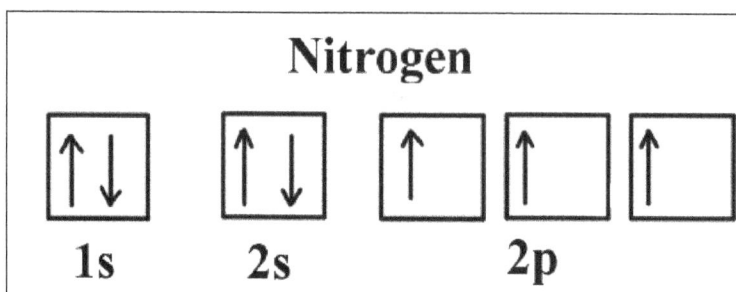

The p orbitals are half-filled; there are three electrons and three p orbitals. This is because the three electrons in the 2p subshell will fill all the empty orbitals first before pairing with electrons in them.

Keep in mind that elemental nitrogen is found in nature typically as dinitrogen, N_2, which requires molecular orbitals instead of atomic orbitals.

Example: Oxygen Atoms.

Next, consider oxygen (Z = 8) atom, the element after nitrogen in the same period; its electron configuration is: $1s^2 2s^2 2p^4$:

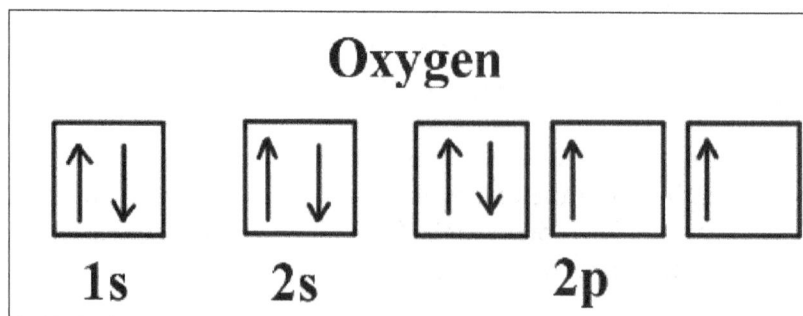

Oxygen has one more electron than nitrogen; as the orbitals are all half-filled, the new electron must pair up. Keep in mind that elemental oxygen is found in nature typically as dioxygen, O_2O_2, which has molecular orbitals instead of atomic orbitals.

Hund's Rule Explained

According to the first rule, electrons always enter an empty orbital before they pair up. Electrons are negatively charged and, as a result, they repel each other. Electrons tend to minimize repulsion by occupying their own orbitals, rather than sharing an orbital with another electron. Furthermore, quantum-mechanical calculations have shown that the electrons in singly occupied orbitals are less effectively screened or shielded from the nucleus.

For the second rule, unpaired electrons in singly occupied orbitals have the same spins. Technically speaking, the first electron in a sublevel could be either "spin-up" or "spin-down". Once the spin of the first electron in a sublevel is chosen, however, the spins of all of the other electrons in that sublevel depend on that first spin. To avoid confusion, scientists typically draw the first electron, and any other unpaired electron, in an orbital as "spin-up".

Example: Carbon and Oxygen.

Consider the electron configuration for carbon atoms: $1s^2 2s^2 2p^2$: The two 2s electrons will occupy the same orbital, whereas the two 2p electrons will be in different orbital (and aligned the same direction) in accordance with Hund's rule.

Consider also the electron configuration of oxygen. Oxygen has 8 electrons. The electron configuration can be written as $1s^2 2s^2 2p^4$. To draw the orbital diagram, begin with the following observations: the first two electrons will pair up in the 1s orbital; the next two electrons will pair up in the 2s orbital. That leaves 4 electrons, which must be placed in the 2p orbitals. According to Hund's rule, all orbitals will be singly occupied before any is doubly occupied. Therefore, two p orbital get one electron and one will have two electrons. Hund's rule also stipulates that all of the unpaired electrons must have the same spin. In keeping with convention, the unpaired electrons are drawn as "spin-up".

Purpose of Electron Configurations

When atoms come into contact with one another, it is the outermost electrons of these atoms, or valence shell, that will interact first. An atom is least stable (and therefore most reactive) when its valence shell is not full. The valence electrons are largely responsible for an element's chemical behavior. Elements that have the same number of valence electrons often have similar chemical properties.

Electron configurations can also predict stability. An atom is most stable (and therefore unreactive) when all its orbitals are full. The most stable configurations are the ones that have full energy levels. These configurations occur in the noble gases. The noble gases are very stable elements that do not react easily with any other elements. Electron configurations can assist in making predictions about the ways in which certain elements will react, and the chemical compounds or molecules that different elements will form.

Hund's Rule of Maximum Multiplicity

Hund's rule of maximum multiplicity states, that in filling p, d or f orbitals, as many unpaired electrons as possible are placed before pairing of electrons with opposite spin is allowed. Pairing of electrons requires energy. Therefore no pairing occurs until all orbitals of a given sub-level are half filled. This is known as Hund's rule of maximum multiplicity. It states that when electrons enter sub-levels of fixed (n+1) values, available orbitals are singly occupied.

Table: Representation of arrangements of electrons.

Atomic Number	Element	1S	2S	2P$_s$	2P$_Y$	2P$_Z$	Number of unpaired electrons
1	H	↑					1
2	He	↑↓					0
3	Li	↑↓	↑				1
4	Be	↑↓	↑↓				0
5	B	↑↓	↑↓	↑	↑		1
6	C	↑↓	↑↓	↑	↑		2
7	N	↑↓	↑↓	↑	↑	↑	3
8	O	↑↓	↑↓	↑↓	↑	↑	2
9	F	↑↓	↑↓	↑↓	↑↓	↑	1
10	Ne	↑↓	↑↓	↑↓	↑↓	↑↓	0

Thus, if three electrons are to be filled in the p- level of any shell, one each will go into each of the three (p_x, p_y, p_z) orbitals. The fourth electron entering the p-level will go to p_x orbital which now will have two electrons with opposite spins and said to be paired. The unpaired electrons play an important part in the formation of bonds.

Aufbau Principle

The Aufbau principle dictates the manner in which electrons are filled in the atomic orbitals of an atom in its ground state. It states that electrons are filled into atomic orbitals in the increasing order of orbital energy level. According to the Aufbau principle, the available atomic orbitals with the lowest energy levels are occupied before those with higher energy levels.

A diagram illustrating the order in which atomic orbitals are filled is provided below. Here, 'n' refers to the principal quantum number and 'l' is the azimuthal quantum number.

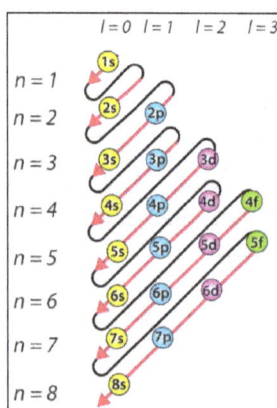

The Aufbau principle can be used to understand the location of electrons in an atom and their corresponding energy levels. For example, carbon has 6 electrons and its electronic configuration is $1s^2 2s^2 2p^2$.

It is important to note that each orbital can hold a maximum of two electrons (as per the Pauli exclusion principle). Also, the manner in which electrons are filled into orbitals in a single subshell must follow Hund's rule, i.e. every orbital in a given subshell must be singly occupied by electrons before any two electrons pair up in an orbital.

Salient Features of the Aufbau Principle

- According to the Aufbau principle, electrons first occupy those orbitals whose energy is the lowest. This implies that the electrons enter the orbitals having higher energies only when orbitals with lower energies have been completely filled.

- The order in which the energy of orbitals increases can be determined with the help of the (n+l) rule, where the sum of the principal and azimuthal quantum numbers determines the energy level of the orbital.

- Lower (n+l) values correspond to lower orbital energies. If two orbitals share equal (n+l) values, the orbital with the lower n value is said to have lower energy associated with it.

- The order in which the orbitals are filled with electrons is: 1s, 2s, 2p, 3s, 3p, 4s, 3d, 4p, 5s, 4d, 5p, 6s, 4f, 5d, 6p, 7s, 5f, 6d, 7p, and so on.

Exceptions

The electron configuration of chromium is $[Ar]3d^5 4s^1$ and not $[Ar]3d^4 4s^2$ (as suggested by the Aufbau principle). This exception is attributed to several factors such as the increased stability provided by half-filled subshells and the relatively low energy gap between the 3d and the 4s subshells.

The energy gap between the different subshells is illustrated below.

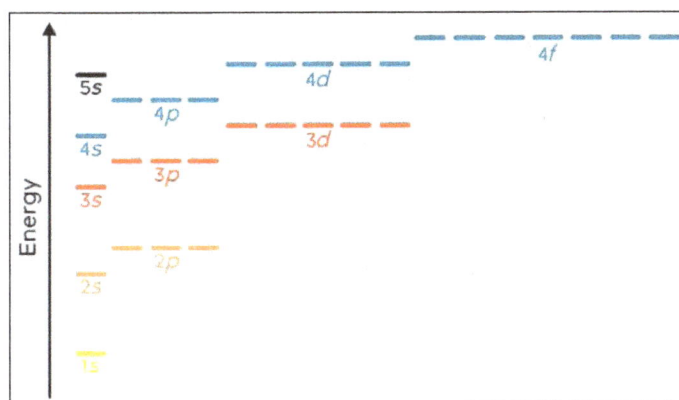

Half filled subshells feature lower electron-electron repulsions in the orbitals, thereby increasing the stability. Similarly, completely filled subshells also increase the stability of the atom. Therefore, the electron configurations of some atoms disobey the Aufbau principle (depending on the energy gap between the orbitals).

For example, copper is another exception to this principle with an electronic configuration corresponding to $[Ar]3d^{10} 4s^1$. This can be explained by the stability provided by a completely filled 3d subshell.

Electronic Configuration using the Aufbau Principle

Writing the Electron Configuration of Sulfur:

- The atomic number of sulfur is 16, implying that it holds a total of 16 electrons.

- As per the Aufbau principle, two of these electrons are present in the 1s subshell, eight of them are present in the 2s and 2p subshell, and the remaining are distributed into the 3s and 3p subshells.

- Therefore, the electron configuration of sulfur can be written as $1s^2 2s^2 2p^2 3s^2 3p^6$.

Writing the Electron Configuration of Nitrogen:

- The element nitrogen has 7 electrons (since its atomic number is 7).

- The electrons are filled into the 1s, 2s, and 2p orbitals.

- The electron configuration of nitrogen can be written as $1s^2 2s^2 2p^3$

References

- Quantum-Chemistry, Pr-Ro: chemistryexplained.com, Retrieved 4 June, 2019

- Jensen, Frank (2007). Introduction to Computational Chemistry. Chichester, England: John Wiley and Sons. Pp. 80–81. ISBN 978-0-470-01187-4

- Applications-quantum-chemistry-chemical-structure-and, chemical-informatics-research-group: nist.gov, Retrieved 5 July, 2019

- Leach, Dr Andrew (2001-01-30). Molecular Modelling: Principles and Applications (2 ed.). Harlow: Prentice Hall. ISBN 9780582382107

- Hund's-Rules, Electronic-Configurations, Electronic-Structure-of-Atoms-and-Molecules, Supplemental-Modules-(Physical-and-Theoretical-Chemistry), Physical-and-Theoretical-Chemistry-Textbook-Maps, Book-shelves: chem.libretexts.org, Retrieved 6 August, 2019

- Werner, H-J; Manby, F. R.; Knowles, P. J. (2003). "Fast linear scaling second-order Møller–Plesset perturbation theory (MP2) using local and density fitting approximations". Journal of Chemical Physics. 118 (18): 8149–8161. Bibcode:2003JChPh.118.8149W. doi:10.1063/1.1564816

- Hund-s-rule-of-maximum-multiplicity-2686: brainkart.com, Retrieved 7 January, 2019

- Aufbau-principle, chemistry: byjus.com, Retrieved 8 February, 2019

2
Many-electron Atoms

Many-electron atom refers to an atom which has more than 1 electron. It includes the aspects of spin quantum number, azimuthal quantum number, magnetic quantum number, angular momentum of electrons, etc. The topics elaborated in this chapter will help in gaining a better perspective about these aspects related to many-electron atoms.

The quantum mechanical model allowed us to determine the energies of the hydrogen atomic orbitals; now we would like to extend this to describe the electronic structure of every element in the Periodic Table. The process of describing each atom's electronic structure consists, essentially, of beginning with hydrogen and adding one proton and one electron at a time to create the next heavier element in the table; however, interactions between electrons make this process a bit more complicated than it sounds.

All stable nuclei other than hydrogen also contain one or more neutrons. Because neutrons have no electrical charge, however, they can be ignored. Before demonstrating how to do this, however, we must introduce the concept of electron spin and the Pauli principle.

Orbitals and their Energies

Unlike in hydrogen-like atoms with only one electron, in multielectron atoms the values of quantum numbers n and l determine the energies of an orbital. The energies of the different orbitals for a typical multielectron atom are shown in figure. Within a given principal shell of a multielectron atom, the orbital energies increase with increasing l. An ns orbital always lies below the corresponding np orbital, which in turn lies below the nd orbital.

These energy differences are caused by the effects of shielding and penetration, the extent to which a given orbital lies inside other filled orbitals. For example, an electron in the 2s orbital penetrates inside a filled 1s orbital more than an electron in a 2p orbital does.

Since electrons, all being negatively charged, repel each other, an electron closer to the nucleus partially shields an electron farther from the nucleus from the attractive effect of the positively charged nucleus. Hence in an atom with a filled 1s orbital, the effective nuclear charge (Z_{eff}) experienced by a 2s electron is greater than the Z_{eff} experienced by a 2p electron. Consequently, the 2s electron is more tightly bound to the nucleus and has a lower energy, consistent with the order of energies shown in figure.

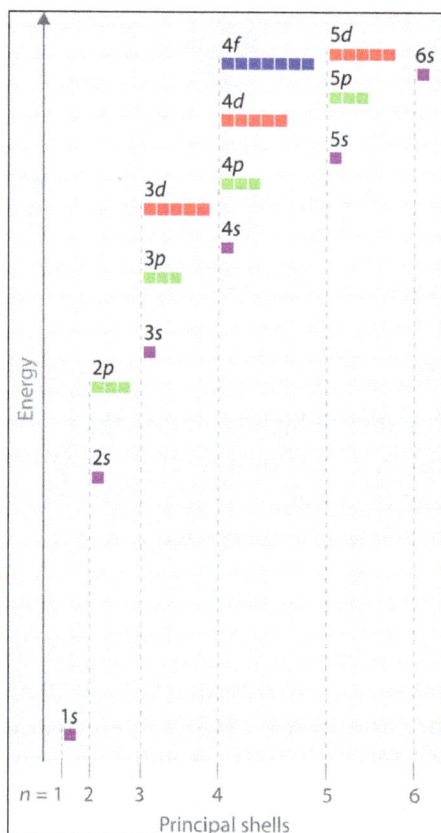

Orbital Energy Level Diagram for a Typical Multielectron Atom.

Due to electron shielding, Z_{eff} increases more rapidly going across a row of the periodic table than going down a column.

Notice in figure that the difference in energies between subshells can be so large that the energies of orbitals from different principal shells can become approximately equal. For example, the energy of the 3d orbitals in most atoms is actually between the energies of the 4s and the 4 porbitals.

Electron Spin: The Fourth Quantum Number

When scientists analyzed the emission and absorption spectra of the elements more closely, they saw that for elements having more than one electron, nearly all the lines in the spectra were actually *pairs* of very closely spaced lines. Because each line represents an energy level available to electrons in the atom, there are twice as many energy levels available as would be predicted solely based on the quantum numbers n, l, and m_l. Scientists also discovered that applying a magnetic field caused the lines in the pairs to split farther apart. In, two graduate students in physics in the Netherlands, George Uhlenbeck and Samuel Goudsmit, proposed that the splittings were caused by an electron spinning about its axis, much as Earth spins about its axis. When an electrically charged object spins, it produces a magnetic moment parallel to the axis of rotation, making it behave like a magnet. Although the electron cannot be viewed solely as a particle, spinning or otherwise, it is indisputable that it does have a magnetic moment. This magnetic moment is called electron spin.

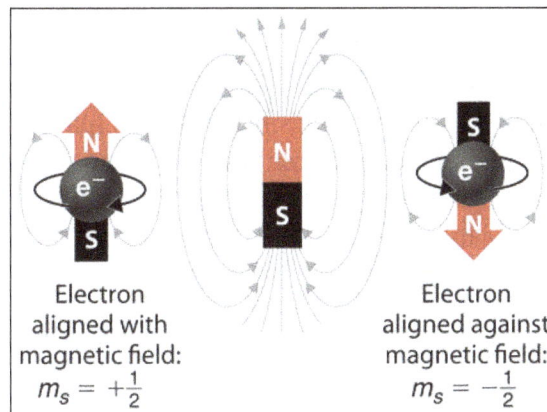

Electron Spin. In a magnetic field, an electron has two possible orientations with different energies, one with spin up, aligned with the magnetic field, and one with spin down, aligned against it. All other orientations are forbidden.

In an external magnetic field, the electron has two possible orientations These are described by a fourth quantum number (m_s), which for any electron can have only two possible values, designated $+\frac{1}{2}$ (up) and $-\frac{1}{2}$ (down) to indicate that the two orientations are opposites; the subscript s is for spin. An electron behaves like a magnet that has one of two possible orientations, aligned either with the magnetic field or against it.

Independent Electron Approximation

The independent electron approximation is used in both the free electron model and the nearly-free electron model. In this approximation we do not consider electron-electron interaction in a crystal. It is more difficult to treat electron-electron interactions than ion-electron interactions because:

- We are not aware of the wavefunctions of every electron.

- The potential due to electron-electron interactions is not periodic.

- We need to consider the dynamics of all of the electrons at once.

Electron-electron interactions are often weaker than ion-electron interactions due to the following:

- Electrons with parallel spins stay away from each other due to the Pauli exclusion principle.

- Electrons with opposite spins stay away from each other in order to have the least energy for the system.

One major effect of electron-electron interactions is that electrons distribute around the ions so that they screen the ions in the lattice from other electrons.

Electron-electron interactions may be very important for certain properties in materials. For example, the theory covering much of superconductivity is BCS theory, in which the attraction of pairs of electrons to each other, termed "Cooper pairs", is the mechanism behind superconductivity.

Pauli Exclusion Principle

The Pauli exclusion principle is the quantum mechanical principle which states that two or more identical fermions (particles with half-integer spin) cannot occupy the same quantum state within a quantum system simultaneously. This principle was formulated by Austrian physicist Wolfgang Pauli in 1925 for electrons, and later extended to all fermions with his spin statistics theorem of 1940.

In the case of electrons in atoms, it can be stated as follows: it is impossible for two electrons of a poly-electron atom to have the same values of the four quantum numbers: n, the principal quantum number, ℓ, the Azimuthal quantum number, m_ℓ, the magnetic quantum number, and m_s, the spin quantum number. For example, if two electrons reside in the same orbital, then their n, ℓ, and m_ℓ values are the same, therefore their m_s must be different, and thus the electrons must have opposite half-integer spin projections of 1/2 and −1/2.

Wolfgang Pauli formulated the law stating that no two electrons can have the same set of quantum numbers.

Particles with an integer spin, or bosons, are not subject to the Pauli exclusion principle: any number of identical bosons can occupy the same quantum state, as with, for instance, photons produced by a laser or atoms in a Bose Einstein condensate.

A more rigorous statement is that with respect to exchange of two identical particles the total wave function is antisymmetric for fermions, and symmetric for bosons. This means that if the space *and* spin co-ordinates of two identical particles are interchanged, then the wave function changes its sign for fermions and does not change for bosons.

The Pauli exclusion principle describes the behavior of all fermions (particles with "half-integer spin"), while bosons (particles with "integer spin") are subject to other principles. Fermions include elementary particles such as quarks, electrons and neutrinos. Additionally, baryons such as protons and neutrons (subatomic particles composed from three quarks) and some atoms (such as helium-3) are fermions, and are therefore described by the Pauli exclusion principle as well. Atoms can have different overall "spin", which determines whether they are fermions or bosons - for example helium-3 has spin 1/2 and is therefore a fermion, in contrast to helium-4 which has spin 0 and is a boson. As such, the Pauli exclusion principle underpins many properties of everyday matter, from its large-scale stability, to the chemical behavior of atoms.

"Half-integer spin" means that the intrinsic angular momentum value of fermions is (reduced Planck's constant) times a half-integer (1/2, 3/2, 5/2, etc).. In the theory of quantum mechanics fermions are described by antisymmetric states. In contrast, particles with integer spin (called bosons) have symmetric wave functions; unlike fermions they may share the same quantum states. Bosons include the photon, the Cooper pairs which are responsible for superconductivity, and the W and Z bosons. (Fermions take their name from the Fermi–Dirac statistical distribution that they obey, and bosons from their Bose Einstein distribution).

Connection to Quantum State Symmetry

The Pauli exclusion principle with a single-valued many-particle wavefunction is equivalent to requiring the wavefunction to be antisymmetric with respect to exchange. An antisymmetric two-particle state is represented as a sum of states in which one particle is in state $|x\rangle$ and the other in state $|y\rangle$, and is given by:

$$|\psi\rangle = \sum_{x,y} A(x,y)|x,y\rangle,$$

and antisymmetry under exchange means that $A(x,y) = -A(y,x)$. This implies $A(x,y) = 0$ when $x = y$, which is Pauli exclusion. It is true in any basis since local changes of basis keep antisymmetric matrices antisymmetric.

Conversely, if the diagonal quantities $A(x,x)$ are zero *in every basis*, then the wavefunction component:

$$A(x,y) = \langle \psi | x, y \rangle = \langle \psi | (|x\rangle \otimes | y\rangle),$$

is necessarily antisymmetric. To prove it, consider the matrix element:

$$\langle \psi | ((|x\rangle + | y\rangle) \otimes (|x\rangle + | y\rangle)).$$

This is zero, because the two particles have zero probability to both be in the superposition state $|x\rangle + |y\rangle$. But this is equal to:

$$\langle \psi | x, x \rangle + \langle \psi | x, y \rangle + \langle \psi | y, x \rangle + \langle \psi | y, y \rangle.$$

The first and last terms are diagonal elements and are zero, and the whole sum is equal to zero. So the wavefunction matrix elements obey:

$$\langle \psi | x, y \rangle + \langle \psi | y, x \rangle = 0,$$

or,

$$A(x,y) = -A(y,x).$$

Advanced Quantum Theory

According to the spin–statistics theorem, particles with integer spin occupy symmetric quantum states, and particles with half-integer spin occupy antisymmetric states; furthermore, only integer or half-integer values of spin are allowed by the principles of quantum mechanics. In relativistic quantum field theory, the Pauli principle follows from applying a rotation operator in imaginary time to particles of half-integer spin.

In one dimension, bosons, as well as fermions, can obey the exclusion principle. A one-dimensional Bose gas with delta-function repulsive interactions of infinite strength is equivalent to a gas of free fermions. The reason for this is that, in one dimension, exchange of particles requires that they pass through each other; for infinitely strong repulsion this cannot happen. This model is described by a quantum nonlinear Schrödinger equation. In momentum space the exclusion principle is valid also for finite repulsion in a Bose gas with delta-function interactions, as well as for interacting spins and Hubbard model in one dimension, and for other models solvable by Bethe ansatz. The ground state in models solvable by Bethe ansatz is a Fermi sphere.

Consequences

Atoms

The Pauli exclusion principle helps explain a wide variety of physical phenomena. One particularly important consequence of the principle is the elaborate electron shell structure of atoms and the way atoms share electrons, explaining the variety of chemical elements and their chemical combinations. An electrically neutral atom contains bound electrons equal in number to the protons in the nucleus. Electrons, being fermions, cannot occupy the same quantum state as other electrons, so electrons have to "stack" within an atom, i.e. have different spins while at the same electron orbital as described below.

An example is the neutral helium atom, which has two bound electrons, both of which can occupy the lowest-energy ($1s$) states by acquiring opposite spin; as spin is part of the quantum state of the electron, the two electrons are in different quantum states and do not violate the Pauli principle. However, the spin can take only two different values (eigenvalues). In a lithium atom, with three bound electrons, the third electron cannot reside in a 1s state, and must occupy one of the higher-energy 2s states instead. Similarly, successively larger elements must have shells of successively higher energy. The chemical properties of an element largely depend on the number of electrons in the outermost shell; atoms with different numbers of occupied electron shells but the same number of electrons in the outermost shell have similar properties, which gives rise to the periodic table of the elements.

Solid State Properties

In conductors and semiconductors, there are very large numbers of molecular orbitals which effectively form a continuous band structure of energy levels. In strong conductors (metals) electrons are so degenerate that they cannot even contribute much to the thermal capacity of a metal. Many mechanical, electrical, magnetic, optical and chemical properties of solids are the direct consequence of Pauli exclusion.

Stability of Matter

The stability of each electron state in an atom is described by the quantum theory of the atom, which shows that close approach of an electron to the nucleus necessarily increases the electron's kinetic energy, an application of the uncertainty principle of Heisenberg. However, stability of large systems with many electrons and many nucleons is a different matter and requires the Pauli exclusion principle.

It has been shown that the Pauli exclusion principle is responsible for the fact that ordinary bulk matter is stable and occupies volume. This suggestion was first made in 1931 by Paul Ehrenfest, who pointed out that the electrons of each atom cannot all fall into the lowest-energy orbital and must occupy successively larger shells. Atoms therefore occupy a volume and cannot be squeezed too closely together.

A more rigorous proof was provided in 1967 by Freeman Dyson and Andrew Lenard, who considered the balance of attractive (electron–nuclear) and repulsive (electron–electron and nuclear–nuclear) forces and showed that ordinary matter would collapse and occupy a much smaller volume without the Pauli principle.

The consequence of the Pauli principle here is that electrons of the same spin are kept apart by a repulsive exchange interaction, which is a short-range effect, acting simultaneously with the long-range electrostatic or Coulombic force. This effect is partly responsible for the everyday observation in the macroscopic world that two solid objects cannot be in the same place at the same time.

Astrophysics

Freeman Dyson and Andrew Lenard did not consider the extreme magnetic or gravitational forces that occur in some astronomical objects. In 1995 Elliott Lieb and coworkers showed that the Pauli principle still leads to stability in intense magnetic fields such as in neutron stars, although at a much higher density than in ordinary matter. It is a consequence of general relativity that, in sufficiently intense gravitational fields, matter collapses to form a black hole.

Astronomy provides a spectacular demonstration of the effect of the Pauli principle, in the form of white dwarf and neutron stars. In both bodies, atomic structure is disrupted by extreme pressure, but the stars are held in hydrostatic equilibrium by degeneracy pressure, also known as Fermi pressure. This exotic form of matter is known as degenerate matter. The immense gravitational force of a star's mass is normally held in equilibrium by thermal pressure caused by heat produced in thermonuclear fusion in the star's core. In white dwarfs, which do not undergo nuclear fusion, an opposing force to gravity is provided by electron degeneracy pressure. In neutron stars, subject to even stronger gravitational forces, electrons have merged with protons to form neutrons. Neutrons are capable of producing an even higher degeneracy pressure, neutron degeneracy pressure, albeit over a shorter range. This can stabilize neutron stars from further collapse, but at a smaller size and higher density than a white dwarf. Neutron stars are the most "rigid" objects known; their Young modulus (or more accurately, bulk modulus) is 20 orders of magnitude larger than that of diamond. However, even this enormous rigidity can be overcome by the gravitational field of a massive star or by the pressure of a supernova, leading to the formation of a black hole.

Spin Quantum Number

The spin quantum number is a quantum number that parameterizes the intrinsic angular momentum (or spin angular momentum, or simply spin) of a given particle. The spin quantum number is the fourth of a set of quantum numbers (the principal quantum number, the azimuthal quantum number, the magnetic quantum number, and the spin quantum number), which completely describe the quantum state of an electron. It is designated by the letter s. It describes the energy, shape and orientation of orbitals. The name comes from a physical spinning (denoted by the letter s) about an axis that was proposed by Uhlenbeck and Goudsmit. However this simplistic picture was quickly realized to be physically impossible and replaced by a more abstract quantum-mechanical description.

Derivation

As a solution for a certain partial differential equation, the quantized angular momentum (see angular momentum quantum number) can be written as:

$$\|s\| = \sqrt{s(s+1)}\,\hbar ,$$

where,

s is the quantized spin vector.

$\|s\|$ is the norm of the spin vector.

S is the spin quantum number associated with the spin angular momentum.

$\hbar$ is the reduced Planck constant.

Given an arbitrary direction z (usually determined by an external magnetic field) the spin z-projection is given by:

$$s_z = m_s \hbar ,$$

where m_s is the secondary spin quantum number, ranging from −s to +s in steps of one. This generates $2s + 1$ different values of m_s.

The allowed values for s are non-negative integers or half-integers. Fermions (such as the electron, proton or neutron) have half-integer values, whereas bosons (e.g., photon, mesons) have integer spin values.

Algebra

The algebraic theory of spin is a carbon copy of the angular momentum in quantum mechanics theory. First of all, spin satisfies the fundamental commutation relation:

$$[S_i, S_j] = i\hbar \partial_{ijk} S_k ,$$
$$\left[S_i, S^2\right] = 0$$

where ϵ_{ijk} is the (antisymmetric) Levi-Civita symbol. This means that it is impossible to know two coordinates of the spin at the same time because of the restriction of the uncertainty principle.

Next, the eigenvectors of S^2 and S_z satisfy:

$$S^2 \,|\, s, m_s \rangle = \hbar^2 s(s+1) \,|\, s, m_s \rangle$$
$$S_z \,|\, s, m_s \rangle = \hbar m_s \,|\, s, m_s \rangle$$
$$S_\pm \,|\, s, m_s \rangle = \hbar \sqrt{s(s+1) - m_s(m_s \pm 1)} \,|\, s, m_s \pm 1 \rangle$$

where $S_\pm = S_x \pm i S_y$ are the creation and annihilation (or "raising" and "lowering" or "up" and "down") operators.

Electron Spin

The spin angular momentum is characterized by a quantum number; s = 1/2 specifically for electrons. In a way analogous to other quantized angular momenta, L, it is possible to obtain an expression for the total spin angular momentum:

$$S = \hbar \sqrt{\frac{1}{2}\left(\frac{1}{2}+1\right)} = \frac{\sqrt{3}}{2}\hbar \,,$$

where,

$\hbar$ is the reduced Planck constant.

The hydrogen spectra fine structure is observed as a doublet corresponding to two possibilities for the z-component of the angular momentum, where for any given direction z:

$$\mathbf{S}_z = \pm\frac{1}{2}\hbar \,,$$

whose solution has only two possible z-components for the electron. In the electron, the two different spin orientations are sometimes called "spin-up" or "spin-down".

The spin property of an electron would give rise to magnetic moment, which was a requisite for the fourth quantum number. The electron spin magnetic moment is given by the formula:

$$\mathbf{\imath}_s = -\frac{e}{2m}gS \,,$$

where,

e is the charge of the electron.

g is the Landé g-factor.

and by the equation:

$$\mathbf{\imath}_z = \pm\frac{1}{2}g\mu_B \,,$$

where μ_B is the Bohr magneton.

When atoms have even numbers of electrons the spin of each electron in each orbital has opposing orientation to that of its immediate neighbor(s). However, many atoms have an odd number of electrons or an arrangement of electrons in which there is an unequal number of "spin-up" and "spin-down" orientations. These atoms or electrons are said to have unpaired spins that are detected in electron spin resonance.

Detection of Spin

When lines of the hydrogen spectrum are examined at very high resolution, they are found to be closely spaced doublets. This splitting is called fine structure, and was one of the first experimental evidences for electron spin. The direct observation of the electron's intrinsic angular momentum was achieved in the Stern–Gerlach experiment.

The Stern–Gerlach Experiment

The theory of spatial quantization of the spin moment of the momentum of electrons of atoms situated in the magnetic field needed to be proved experimentally. In 1920 (two years before the theoretical description of the spin was created) Otto Stern and Walter Gerlach observed it in the experiment they conducted.

Silver atoms were evaporated using an electric furnace in a vacuum. Using thin slits, the atoms were guided into a flat beam and the beam sent through an in-homogeneous magnetic field before colliding with a metallic plate. The laws of classical physics predict that the collection of condensed silver atoms on the plate should form a thin solid line in the same shape as the original beam. However, the in-homogeneous magnetic field caused the beam to split in two separate directions, creating two lines on the metallic plate.

The phenomenon can be explained with the spatial quantization of the spin moment of momentum. In atoms the electrons are paired such that one spins upward and one downward, neutralizing the effect of their spin on the action of the atom as a whole. But in the valence shell of silver atoms, there is a single electron whose spin remains unbalanced.

The unbalanced spin creates spin magnetic moment, making the electron act like a very small magnet. As the atoms pass through the in-homogeneous magnetic field, the force moment in the magnetic field influences the electron's dipole until its position matches the direction of the stronger field. The atom would then be pulled toward or away from the stronger magnetic field a specific amount, depending on the value of the valence electron's spin. When the spin of the electron is $+1/2$ the atom moves away from the stronger field, and when the spin is $-1/2$ the atom moves toward it. Thus the beam of silver atoms is split while traveling through the in-homogeneous magnetic field, according to the spin of each atom's valence electron.

In 1927 Phipps and Taylor conducted a similar experiment, using atoms of hydrogen with similar results. Later scientists conducted experiments using other atoms that have only one electron in their valence shell: (copper, gold, sodium, potassium). Every time there were two lines formed on the metallic plate.

The atomic nucleus also may have spin, but protons and neutrons are much heavier than electrons (about 1836 times), and the magnetic dipole moment is inversely proportional to the mass. So the nuclear magnetic dipole momentum is much smaller than that of the whole atom. This small magnetic dipole was later measured by Stern, Frisch and Easterman.

Energy Levels from The Dirac Equation

In 1928, Paul Dirac developed a relativistic wave equation, now termed the Dirac equation, which predicted the spin magnetic moment correctly, and at the same time treated the electron as a point-like particle. Solving the Dirac equation for the energy levels of an electron in the hydrogen atom, all four quantum numbers including s occurred naturally and agreed well with experiment.

Total Spin of An Atom or Molecule

For some atoms the spins of several unpaired electrons (s_1, s_2, ..). are coupled to form a *total spin* quantum number S. This occurs especially in light atoms (or in molecules formed only of light atoms) when spin-orbit coupling is weak compared to the coupling between spins or the coupling between orbital angular momenta, a situation known as LS coupling because L and S are constants of motion. Here L is the total orbital angular momentum quantum number.

For atoms with a well-defined S, the multiplicity of a state is defined as (2S+1). This is equal to the number of different possible values of the total (orbital plus spin) angular momentum J for a given (L, S) combination, provided that S ≤ L (the typical case). For example, if S = 1, there are three states which form a triplet. The eigenvalues of S_z for these three states are +1ℏ, 0 and -1ℏ. The term symbol of an atomic state indicates its values of L, S, and J.

Principal Quantum Number

The principal quantum number (symbolized n) is one of four quantum numbers assigned to all electrons in an atom to describe that electron's state. Its values are natural numbers (from 1) making it a discrete variable.

Apart from the principal quantum number, the other quantum numbers for bound electrons are the azimuthal quantum number ℓ, the magnetic quantum number m_l, and the spin quantum number s.

As n increases, the electron is also at a higher energy and is, therefore, less tightly bound to the nucleus. For higher n the electron is, in average, farther from the nucleus. For each value of n there are n accepted ℓ (azimuthal) values ranging from 0 to $n - 1$ inclusively, hence higher-n electron states are more numerous. Accounting for two states of spin, each n-shell can accommodate up to $2n^2$ electrons.

In a simplistic one-electron model described below, the total energy of an electron is a negative inverse quadratic function of the principal quantum number n, leading to degenerate energy levels for each $n > 1$. In more complex systems-those having forces other than the nucleus–electron Coulomb force-these levels split. For multielectron atoms this splitting results in "subshells" parametrized by ℓ. Description of energy levels based on n alone gradually becomes inadequate for atomic numbers starting from 5 (boron) and fails completely on potassium ($Z = 19$) and afterwards.

The principal quantum number was first created for use in the semiclassical Bohr model of the atom, distinguishing between different energy levels. With the development of modern quantum mechanics, the simple Bohr model was replaced with a more complex theory of atomic orbitals. However, the modern theory still requires the principal quantum number.

Derivation

There is a set of quantum numbers associated with the energy states of the atom. The four quantum numbers n, ℓ, m, and s specify the complete and unique quantum state of a single electron in an atom, called its wave function or orbital. Two electrons belonging to the same atom cannot have the same values for all four quantum numbers, due to the Pauli exclusion principle. The wavefunction of the Schrödinger wave equation reduces to the three equations that when solved lead to the first three quantum numbers. Therefore, the equations for the first three quantum numbers are all interrelated. The principal quantum number arose in the solution of the radial part of the wave equation as shown below.

The Schrödinger wave equation describes energy eigenstates with corresponding real numbers E_n and a definite total energy, the value of E_n. The bound state energies of the electron in the hydrogen atom are given by:

$$E_n = \frac{E_1}{n^2} = \frac{-13.6 \text{ eV}}{n^2}, \quad n = 1, 2, 3, \ldots$$

The parameter n can take only positive integer values. The concept of energy levels and notation were taken from the earlier Bohr model of the atom. Schrödinger's equation developed the idea from a flat two-dimensional Bohr atom to the three-dimensional wavefunction model.

In the Bohr model, the allowed orbits were derived from quantized (discrete) values of orbital angular momentum, L according to the equation:

$$\mathbf{L} = n \cdot \hbar = n \cdot \frac{h}{2\pi},$$

where $n = 1, 2, 3, \ldots$ and is called the principal quantum number, and h is Planck's constant. This formula is not correct in quantum mechanics as the angular momentum magnitude is described by the azimuthal quantum number, but the energy levels are accurate and classically they correspond to the sum of potential and kinetic energy of the electron.

The principal quantum number n represents the relative overall energy of each orbital. The energy level of each orbital increases as its distance from the nucleus increases. The sets of orbitals with the same n value are often referred to as an electron shell.

The minimum energy exchanged during any wave–matter interaction is the product of the wave frequency multiplied by Planck's constant. This causes the wave to display particle-like packets of energy called quanta. The difference between energy levels that have different n determine the emission spectrum of the element.

In the notation of the periodic table, the main shells of electrons are labeled:

$$K(n = 1), L(n = 2), M(n = 3), \text{etc.},$$

based on the principal quantum number.

The principal quantum number is related to the radial quantum number, n_r, by:

$$n = n_r + \ell + 1 ,$$

where ℓ is the azimuthal quantum number and n_r is equal to the number of nodes in the radial wavefunction.

The definite total energy for a particle motion in a common Coulomb field and with a discrete spectrum, is given by:

$$E_n = -\frac{Z^2 \hbar^2}{2 m_0 a_B^2 n^2} = -\frac{Z^2 e^4 m_0}{2 \hbar^2 n^2} ,$$

where a_B is the Bohr radius.

This discrete energy spectrum resulted from the solution of the quantum mechanical problem on the electron motion in the Coulomb field, coincides with the spectrum that was obtained with the help application of the Bohr–Sommerfeld quantization rules to the classical equations. The radial quantum number determines the number of nodes of the radial wave function $R(r)$.

Values

In chemistry, values n = 1, 2, 3, 4, 5, 6, 7 are used in relation to the electron shell theory, with expected inclusion of n = 8 (and possibly 9) for yet not accessible period-8 elements. In atomic physics higher n occur for description of excited states. Observations of interstellar medium reveal atomic hydrogen spectral lines involving n on order of hundreds; values up to 766 were detected.

Azimuthal Quantum Number

The azimuthal quantum number is a quantum number for an atomic orbital that determines its orbital angular momentum and describes the shape of the orbital. The azimuthal quantum number is the second of a set of quantum numbers which describe the unique quantum state of an electron (the others being the principal quantum number, the magnetic quantum number, and the spin quantum number). It is also known as the orbital angular momentum quantum number, orbital quantum number or second quantum number, and is symbolized as ℓ.

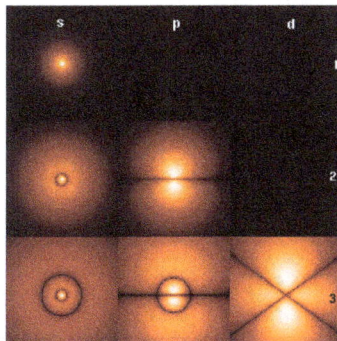

The atomic orbital wavefunctions of a hydrogen atom. The principal quantum number (n) is at the right of each row and the azimuthal quantum number (ℓ) is denoted by letter at top of each column.

Derivation

Connected with the energy states of the atom's electrons are four quantum numbers: n, ℓ, m_ℓ, and m_s. These specify the complete, unique quantum state of a single electron in an atom, and make up its wavefunction or orbital. The wavefunction of the Schrödinger equation reduces to three equations that, when solved, lead to the first three quantum numbers. Therefore, the equations for the first three quantum numbers are all interrelated. The azimuthal quantum number arose in the solution of the polar part of the wave equation as shown below, reliant on the spherical coordinate system, which generally works best with models having some glimpse of spherical symmetry.

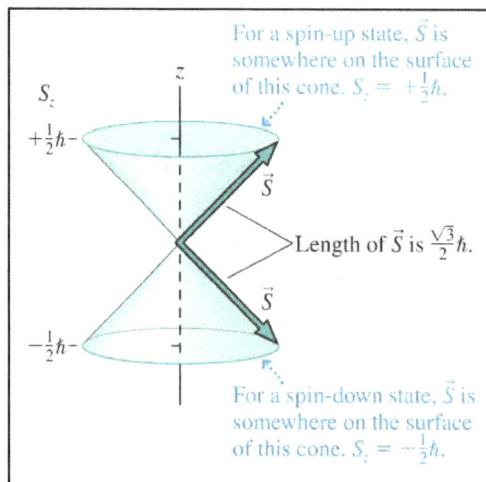

Illustration of quantum mechanical orbital angular momentum.

An atomic electron's angular momentum, L, is related to its quantum number ℓ by the following equation:

$$\mathbf{L}^2\Psi = \hbar^2\ell(\ell+1)\Psi$$

where $\hbar$ is the reduced Planck's constant, $\mathbf{L}^2$ is the orbital angular momentum operator and Ψ is the wavefunction of the electron. The quantum number ℓ is always a non-negative integer: 0, 1, 2, 3, etc. While many introductory textbooks on quantum mechanics will refer to L by itself, L has no real meaning except in its use as the angular momentum operator. When referring to angular momentum, it is better to simply use the quantum number ℓ.

Atomic orbitals have distinctive shapes denoted by letters. In the illustration, the letters s, p, and d (a convention originating in spectroscopy) describe the shape of the atomic orbital.

Their wavefunctions take the form of spherical harmonics, and so are described by Legendre polynomials. The various orbitals relating to different values of ℓ are sometimes called sub-shells, and are referred to by lowercase Latin letters (chosen for historical reasons), as follows:

Quantum Subshells for the Azimuthal Quantum Number				
Azimuthal number (ℓ)	Historical Letter	Maximum Electrons	Historical Name	Shape
0	s	2	sharp	spherical

1	p	6	principal	three dumbbell-shaped polar-aligned orbitals; one lobe on each pole of the x, y, and z (+ and – axes)
2	d	10	diffuse	nine dumbbells and one doughnut (or "unique shape #1" see this picture of spherical harmonics, third row center)
3	f	14	funda-mental	"unique shape #2" (see this picture of spherical harmonics, bottom row center)
4	g	18		
5	h	22		
6	i	26		
The letters after the f sub-shell just follow letter f in alphabetical order except the letter j and those already used.				

Each of the different angular momentum states can take $2(2\ell + 1)$ electrons. This is because the third quantum number m_ℓ (which can be thought of loosely as the quantized projection of the angular momentum vector on the z-axis) runs from $-\ell$ to ℓ in integer units, and so there are $2\ell + 1$ possible states. Each distinct n, ℓ, m_ℓ orbital can be occupied by two electrons with opposing spins (given by the quantum number $m_s = \pm\frac{1}{2}$), giving $2(2\ell + 1)$ electrons overall. Orbitals with higher ℓ than given in the table are perfectly permissible, but these values cover all atoms so far discovered.

For a given value of the principal quantum number n, the possible values of ℓ range from 0 to n – 1; therefore, the n = 1 shell only possesses an s subshell and can only take 2 electrons, the n = 2 shell possesses an s and a p subshell and can take 8 electrons overall, the n = 3 shell possesses s, p, and d subshells and has a maximum of 18 electrons, and so on.

A simplistic one-electron model results in energy levels depending on the principal number alone. In more complex atoms these energy levels split for all n > 1, placing states of higher ℓ above states of lower ℓ. For example, the energy of 2p is higher than of 2s, 3d occurs higher than 3p, which in turn is above 3s, etc. This effect eventually forms the block structure of the periodic table. No known atom possesses an electron having ℓ higher than three (f) in its ground state.

The angular momentum quantum number, ℓ, governs the number of planar nodes going through the nucleus. A planar node can be described in an electromagnetic wave as the midpoint between crest and trough, which has zero magnitude. In an s orbital, no nodes go through the nucleus, therefore the corresponding azimuthal quantum number ℓ takes the value of 0. In a p orbital, one node traverses the nucleus and therefore ℓ has the value of 1. **L** has the value $\sqrt{2}\hbar$.

Depending on the value of n, there is an angular momentum quantum number ℓ and the following series. The wavelengths listed are for a hydrogen atom:

- $n = 1, L = 0$, Lyman series (ultraviolet).

- $n = 2, L = \sqrt{2}\hbar$, Balmer series (visible).

- $n = 3, L = \sqrt{6}\hbar$, Ritz–Paschen series (near infrared).

- $n = 4, L = 2\sqrt{3}\hbar$, Brackett series (short-wavelength infrared).

- $n = 5, L = 2\sqrt{5}\hbar$ Pfund series (mid-wavelength infrared).

Addition of Quantized Angular Momenta

Given a quantized total angular momentum $\vec{j}$ which is the sum of two individual quantized angular momenta $\vec{\ell_1}$ and $\vec{\ell_2}$,

$$\vec{j} = \vec{\ell_1} + \vec{\ell_2},$$

the quantum number j associated with its magnitude can range from $|\ell_1 - \ell_2|$ to $\ell_1 + \ell_2$ in integer steps where ℓ_1 and ℓ_2 are quantum numbers corresponding to the magnitudes of the individual angular momenta.

Total Angular Momentum of an Electron in the Atom

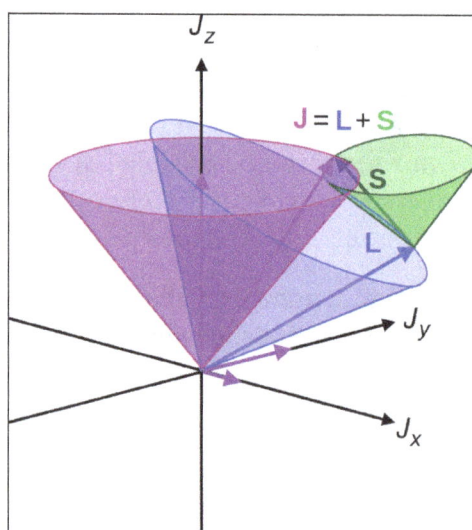

"Vector cones" of total angular momentum J (purple), orbital L (blue), and spin S (green).
The cones arise due to quantum uncertainty between measuring angular momentum
components (see vector model of the atom.

Due to the spin–orbit interaction in the atom, the orbital angular momentum no longer commutes with the Hamiltonian, nor does the spin. These therefore change over time. However the total angular momentum J does commute with the one-electron Hamiltonian and so is constant. J is defined through:

$$\vec{J} = \vec{L} + \vec{S},$$

L being the orbital angular momentum and S the spin. The total angular momentum satisfies the same commutation relations as orbital angular momentum, namely:

$$[J_i, J_j] = i\hbar \epsilon_{ijk} J_k,$$

from which follows:

$$[J_i, J^2] = 0,$$

where J_i stand for J_x, J_y, and J_z.

The quantum numbers describing the system, which are constant over time, are now j and mj, defined through the action of J on the wavefunction Ψ

$$J^2 \psi = \hbar^2 j(j+1)\psi$$
$$J_z \psi = \hbar m_j \psi$$

So that j is related to the norm of the total angular momentum and m_j to its projection along a specified axis. The j number has a particular importance for relativistic quantum chemistry, often featuring in subscript in electron configuration of superheavy elements.

As with any angular momentum in quantum mechanics, the projection of J along other axes cannot be co-defined with J_z, because they do not commute.

Relation between new and Old Quantum Numbers

j and m_j, together with the parity of the quantum state, replace the three quantum numbers ℓ, m_ℓ and m_s (the projection of the spin along the specified axis). The former quantum numbers can be related to the latter.

Furthermore, the eigenvectors of j, s, m_j and parity, which are also eigenvectors of the Hamiltonian, are linear combinations of the eigenvectors of ℓ, s, m_ℓ and m_s.

List of Angular Momentum Quantum Numbers

- Intrinsic (or spin) angular momentum quantum number, or simply spin quantum number.
- Orbital angular momentum quantum number.
- Magnetic quantum number, related to the orbital momentum quantum number.
- Total angular momentum quantum number.

Magnetic Quantum Number

The magnetic quantum number (symbol m_l) is one of four quantum numbers in atomic physics. The set is: principal quantum number, azimuthal quantum number, magnetic quantum number, and spin quantum number. Together, they describe the unique quantum state of an electron. The magnetic quantum number distinguishes the orbitals available within a subshell, and is used to calculate the azimuthal component of the orientation of orbital in space. Electrons in a particular subshell (such as s, p, d, or f) are defined by values of ℓ (0, 1, 2, or 3). The value of m_l can range from $-\ell$ to $+\ell$, inclusive of zero. Thus the s, p, d, and f subshells contain 1, 3, 5, and 7 orbitals each, with values of m within the ranges 0, ±1, ±2, ±3 respectively. Each of these orbitals can accommodate up to two electrons (with opposite spins), forming the basis of the periodic table.

Derivation

There is a set of quantum numbers associated with the energy states of the atom. The four quantum numbers n, ℓ, m, and s specify the complete and unique quantum state of a single electron in an atom called its wavefunction or orbital. The Schrödinger equation for the wavefunction of an atom with one electron is a separable partial differential equation. (This is not the case for the helium atom or other atoms with mutually interacting electrons, which require more sophisticated methods for solution) This means that the wavefunction as expressed in spherical coordinates can be broken down into the product of three functions of the radius, colatitude (or polar) angle, and azimuth:

$$\psi(r,\theta,\phi) = R(r)P(\theta)F(\phi)$$

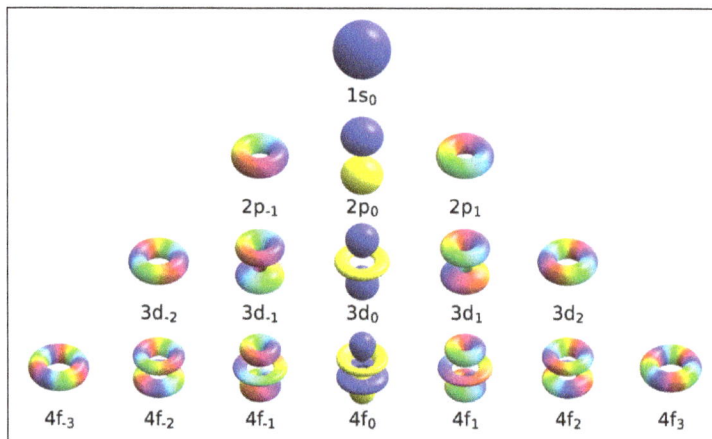

These orbitals have magnetic quantum numbers $m = -\ell, \ldots, \ell$ from left to right in ascending order. The $e^{mi\phi}$ dependence of the azimuthal component can be seen as a color gradient repeating m times around the vertical axis.

The differential equation for F can be solved in the form $F(\phi) = Ae^{\lambda\phi}$. Because values of the azimuth angle ϕ differing by 2π (360 degrees in radians) represent the same position in space, and the overall magnitude of F does not grow with arbitrarily large ϕ as it would for a real exponent, the coefficient λ must be quantized to integer multiples of, producing an imaginary exponent: $\lambda = im$. These integers are the magnetic quantum numbers. The same constant appears in the colatitude equation, where larger values of ℓ^2 tend to decrease the magnitude of m^2, and values of $P(\theta)$. greater than the azimuthal quantum number $P(\theta)$. do not permit any solution for $P(\theta)$.

Relationship between Quantum Numbers			
Orbital	Values	Number of Values for	Electrons per subshell
s	$\ell = 0, m_l = 0$	1	2
p	$\ell = 1, m_l = -1, 0, +1$	3	6
d	$\ell = 2, m_l = -2, -1, 0, +1, +2$	5	10
f	$\ell = 3, m = -3, -2, -1, 0, +1, +2, +3$	7	14
g	$\ell = 4, m = -4, -3, -2, -1, 0, +1, +2, +3, +4$	9	18

Component of Angular Momentum

The axis used for the polar coordinates in this analysis is chosen arbitrarily. The quantum number m refers to the projection of the angular momentum in this arbitrarily-chosen direction, conventionally called the z direction or quantization axis. L_z, the magnitude of the angular momentum in the z direction, is given by the formula:

$$L_z = m\hbar.$$

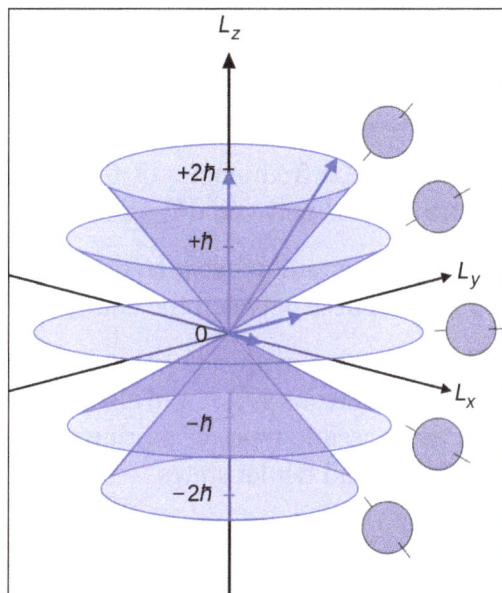

Illustration of quantum mechanical orbital angular momentum. The cones and plane represent possible orientations of the angular momentum vector for $\ell = 2$ and $m = -2, -1, 0, 1, 2$. Even for the extreme values of m, the z component of this vector is less than its total magnitude.

This is a component of the atomic electron's total orbital angular momentum, L, whose magnitude is related to the azimuthal quantum number of its subshell ℓ by the equation:

$$L = \hbar\sqrt{\ell(\ell+1)}$$

where $\hbar$ is the reduced Planck constant. Note this $L = 0$ for $\ell = 0$ and approximates $L = (\ell + 0.5)\hbar$ for high ℓ. It is not possible to measure the angular momentum of the electron in all three axes simultaneously. These properties were first demonstrated in the Stern-Gerlach experiment, by Otto Stern and Walther Gerlach.

The energy of any wave is its frequency multiplied by Planck's constant. The wave displays particle-like packets of energy called quanta. The formula for the quantum number of each quantum state uses Planck's reduced constant, which only allows particular or discrete or quantized energy levels.

Effect in Magnetic Fields

The quantum number m refers, loosely, to the direction of the angular momentum vector. The magnetic quantum number m only affects the electron's energy if it is in a magnetic field because in the absence of one, all spherical harmonics corresponding to the different arbitrary values of m are equivalent. The magnetic quantum number determines the energy shift of an atomic orbital due to an external magnetic field (the Zeeman effect) - hence the name *magnetic* quantum number. However, the actual magnetic dipole moment of an electron in an atomic orbital arrives not only from the electron angular momentum, but also from the electron spin, expressed in the spin quantum number.

Since each electron has a magnetic moment in a magnetic field, it will be subject to a torque which tends to make the vector L parallel to the field, a phenomenon known as Larmor precession.

Singlet and Triplet States

In a hydrogenic (one electron) species, the frequencies of the lines in the atomic absorption or emission spectra corresponds to transitions between different energy levels.

In atoms with more than one electron, interaction between the electrons means that the transition energies are not simply related to differences between the energies of the different energy levels.

The electrons in different orbitals can interact with each other in a range of possible ways, depending on the relative orientations on the electrons spins, and also the nature of the orbitals occupied. In a two electron system, the are two possible types of pairing of the electrons, when in different orbitals, and these are known as singlet and triplet states.

Let us consider Helium:

> The ground state electronic configuration is 1s2, and the first excited state is 1s12s1. However, whilst the electrons must have paired spins when both occupy the 1s orbital, they need not be when one occupies the 1s orbital and the other occupies the 2s orbital.

> Hund's rule states that the configuration with the electron spins parallel is at a lower energy than the configuration with paired spins. When the electron spins are paired, their individual spin momenta cancel each other out, and there is no overall spin.

> There is only one orientation in which this may be achieved, and hence this configuration is known as a singlet state. When the electron spins are parallel, their individual spin momenta add together to give non-zero total spin, and this can be achieved in three ways. Thus, the state is known as a triplet state.

> Hund's rule therefore means that the triplet excited state of He is lower in energy than the singlet excited state of He.

The degree of electron-electron repulsion depends greatly upon the state the two electrons occupy, and the difference between the triplet and singlet states in He is 0.8 eV.

Whilst the spectrum of helium might be more complicated than that of hydrogen due to the splitting of the levels into singlet and triplet states, there are two important simplifications in interpreting the spectrum: the first is that only one electron may be promoted from the ground state, and the second is that no transitions occur from a triplet state to a single state, or vice versa.

This is because the relative orientations of the electron spin cannot change during a transition.

Angular Momentum of Electrons

The simplest classical model of the hydrogen atom is one in which the electron moves in a circular planar orbit about the nucleus as previously discussed and as illustrated in figure. The angular momentum vector M in this figure is shown at an angle q with respect to some arbitrary axis in space. Assuming for the moment that we can somehow physically define such an axis, then in the classical model of the atom there should be an infinite number of values possible for the component of the angular momentum vector along this axis. As the angle between the axis and the vector M varies continuously from 0°, through 90° to 180°, the component of M along the axis would vary correspondingly from M to zero to -M. Thus the quantum mechanical statements regarding the angular momentum of an electron in an atom differ from the classical predictions in two startling ways. First, the magnitude of the angular momentum (the length of the vector M) is restricted to only certain values given by:

$$\sqrt{l(l+1)}\,\frac{h}{2\pi} \qquad l = 0,1,2,...$$

The magnitude of the angular momentum is quantized. Secondly, quantum mechanics states that the component of M along a given axis can assume only $(2l + 1)$ values, rather than the infinite number allowed in the classical model. In terms of the classical model this would imply that when the magnitude of M is $\sqrt{2}\,(h/2\pi)$ (the value when l = 1), there are only three allowed values for q, the angle of inclination of M with respect to a chosen axis.

The angle θ is another example of a physical quantity which in a classical system may assume any value, but which in a quantum system may take on only certain discrete values. You need not accept this result on faith. There is a simple, elegant experiment which illustrates the "quantization" of θ, just as a line spectrum illustrates the quantization of the energy.

If we wish to measure the number of possible values which the component of the angular momentum may exhibit with respect to some axis we must first find some way in which we can physically define a direction or axis in space. To do this we make use of the magnetism exhibited by an electron in an atom. The flow of electrons through a loop of wire (an electric current) produces a magnetic field. At a distance from the ring of wire, large compared to the diameter of the ring, the magnetic field produced by the current appears to be the same as that obtained from a small bar magnet with a north pole and a south pole. Such a small magnet is called a magnetic dipole, i.e., two poles separated by a small distance.

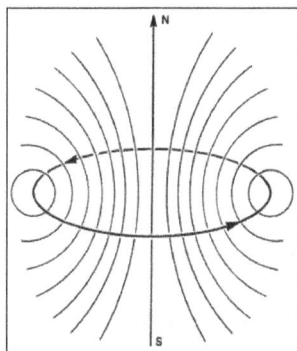

The magnetic field produced by a current in a loop of wire.

The electron is charged and the motion of the electron in an atom could be thought of as generating a small electric current. Associated with this current there should be a small magnetic field. The magnitude of this magnetic field is related to the angular momentum of the electron's motion in roughly the same way that the magnetic field produced by a current in a loop of wire is proportional to the strength of the current flowing in the wire.

The strength of the atomic magnetic dipole is given by m where:

$$\mu = \sqrt{l(l+1)}\, \beta_m$$

Just as there is a fundamental unit of negative charge denoted by e^- so there is a fundamental unit of magnetism at the atomic level denoted by β_m and called the Bohr magneton. From equation $\mu = \sqrt{l(l+1)}\, \beta_m$ we can see that the strength of the magnetic dipole will increase as the angular momentum of the electron increases. This is analogous to increasing the magnetic field by increasing the strength of the current through a circular loop of wire The magnetic dipole, since it has a north and a south pole, will define some direction in space (the magnetic dipole is a vector quantity). The axis of the magnetic dipole in fact coincides with the direction of the angular momentum vector. Experimentally, a collection of atoms behave as though they were a collection of small bar magnets if the electrons in these atoms possess angular momentum. In addition, the axis of the magnet lies along the axis of rotation, i.e., along the angular momentum vector. Thus the magnetism exhibited by the atoms provides an experimental means by which we may study the direction of the angular momentum vector.

If we place the atoms in a magnetic field they will be attracted or repelled by this field, depending on whether or not the atomic magnets are aligned against or with the applied field. The applied magnetic field will determine a direction in space. By measuring the deflection of the atoms in this field we can determine the directions of their magnetic moments and hence of their angular momentum vectors with respect to this applied field. Consider an evacuated tube with a tiny opening at one end through which a stream of atoms may enter. By placing a second small hole in front of the first, inside the tube, we will obtain a narrow beam of atoms which will pass the length of the tube and strike the opposite end. If the atoms possess magnetic moments the path of the beam can be deflected by placing a magnetic field across the tube, perpendicular to the path of the atoms. The magnetic field must be one in which the lines of force diverge thereby exerting an unbalanced force on any magnetic material lying inside the field. This inhomogeneous magnetic field could be obtained through the use of N and S poles of the kind illustrated in figure. The direction of the magnetic field will be taken as the direction of the z-axis.

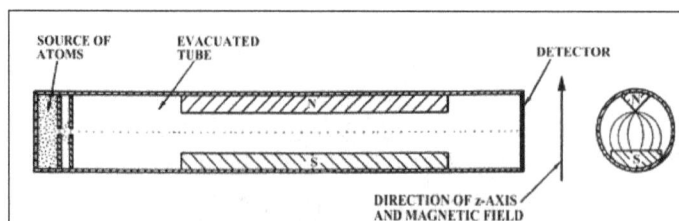

The atomic beam apparatus.

Let us suppose the beam consists of neutral atoms which possess $\sqrt{2}\,(h/2\pi)$ units of electronic angular momentum (the angular momentum quantum number $l = 1$). When no magnetic field is present, the beam of atoms strikes the end wall at a single point in the middle of the detector. What happens when the magnetic field is present? We must assume that before the beam enters the magnetic field, the axes of the atomic magnets are randomly oriented with respect to the z-axis.

According to the concepts of classical mechanics, the beam should spread out along the direction of the magnetic field and produce a line rather than a point at the end of the tube. Actually, the beam is split into three distinct component beams each of equal intensity producing three spots at the end of the tube.

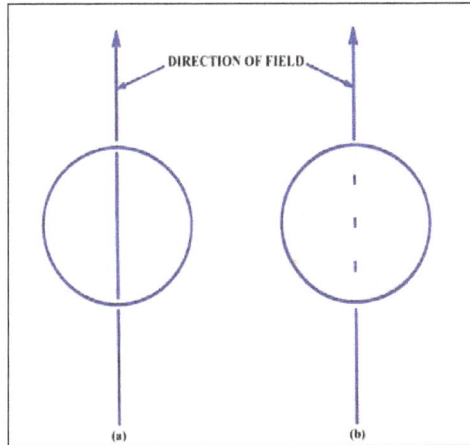

(a) The result of the atomic beam experiment as predicted by classical mechanics,
(b) The observed result of the atomic beam experiment.

The startling results of this experiment can be explained only if we assume that while in the magnetic field each atomic magnet could assume only one of three possible orientations with respect to the applied magnetic field.

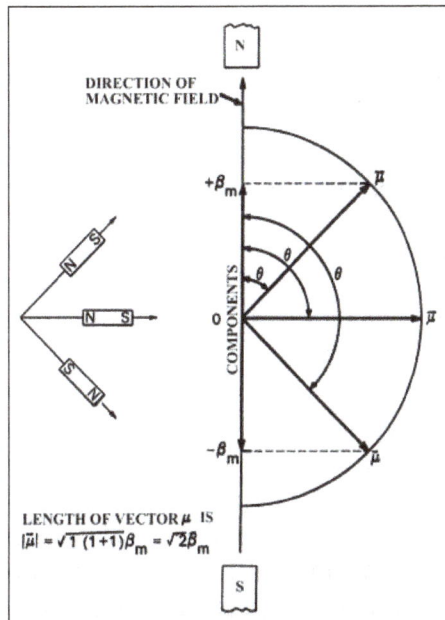

The three possible orientations for the total magnetic moment with
respect to an external magnetic field for an atom with $l = 1$.

The atomic magnets which are aligned perpendicular to the direction of the field are not deflected and will follow a straight path through the tube. The atoms which are attracted upwards must have their magnetic moments oriented as shown. From the known strength of the applied inhomogeneous magnetic field and the measured distance through which the beam has been deflected upwards, we can determine that the component of the magnetic moment lying along the z-axis is only β_m in magnitude rather than the value of $\sqrt{2}\,\beta_m$ This latter value would result if the axis of the atomic magnet was parallel to the z-axis, i.e., the angle $\theta = 0°$. Instead q assumes a value such that the component of the total moment $(\sqrt{l(l+1)}\beta_m)$ lying along the z-axis is just $l\beta_m$. Similarly the beam which is deflected downwards possesses a magnetic moment along the z-axis of $-\beta_m$. or -   . The classical prediction for this experiment assumes that q may equal all values from 0° to 180°, and thus all values (from a maximum of $-\sqrt{2}\,\beta_m$ ($\theta = 0°$) to o ($\theta = 90°$) to $-\sqrt{2}\,\beta_m$ ($\theta = 180°$)) for the component of the atomic moment along the z-axis would be possible. Instead, q is found to equal only those values such that the magnetic moment along the z-axis equals β_m, o and $-\beta_m$.

The angular momentum of the electron determines the magnitude and the direction of the magnetic dipole. (Recall that the vectors for both these quantities lie along the same axis). Thus the number of possible values which the component of the angular momentum vector may assume along a given axis must equal the number of values observed for the component of the magnetic dipole along the same axis. In the present example the values of the angular momentum component are $+1(h/2\pi)$, o and $-1(h/2\pi)$, or since $l=1$ in this case, $+l(h/2\pi)$, o and $-l(h/2)$. In general, it is found that the number of observed values is always $(2l+1)$ the values being:

$$l\left(\frac{h}{2\pi}\right),(l+1)\left(\frac{h}{2\pi}\right),...,0..,-l\left(\frac{h}{2\pi}\right),$$

for the angular momentum:

$$l\beta_m,(l+1)\beta_m,...,0,...,-l\beta_m,$$

for the magnetic dipole. The number governing the magnitude of the component of M and $\bar{\mu}$, ranges from a maximum value of l and decreases in steps of unity to a minimum value of $-l$. This number is the third and final quantum number which determines the motion of an electron in a hydrogen atom. It is given the symbol m and is called the magnetic quantum number.

The angular momentum of an electron in the hydrogen atom is quantized and may assume only those values given by:

$$\sqrt{l(l+1)}\frac{h}{2\pi}, \text{ where } l = 0,1,2,...n-1,$$

Further more, it is an experimental fact that the component of the angular momentum vector along a given axis is limited to $(2l + 1)$ different values, and that the magnitude of this component is quantized and governed by the quantum number m which may assume the values $l, l-1, ...,0,...,-l$. These facts are illustrated in figure for an electron in a d orbital in which $l = 2$.

$$|M|\sqrt{2(2+1)}=\sqrt{6}$$

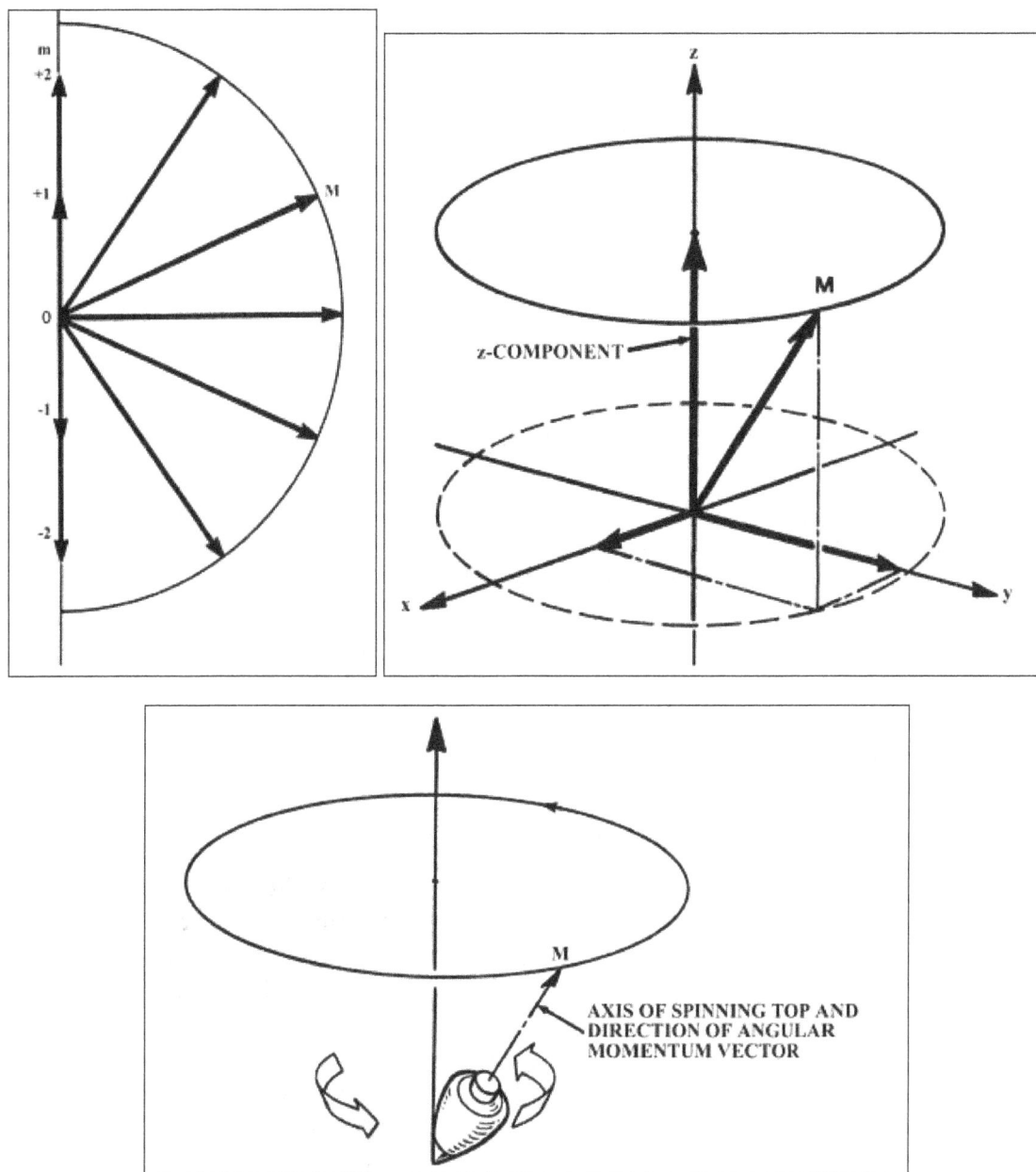

Figure Pictorial representation of the quantum mechanical properties of the angular momentum of a d electron for which l = 2. The z-axis can be along any arbitrary direction in space. Figure shows the possible components which the angular momentum vector (of length $\sqrt{2(2+1)}(h/2\pi)$) may exhibit along an arbitrary axis in space. A d electron may possess any one of these components. There are therefore five states for a d electron, all of which are physically different. Notice that the maximum magnitude allowed for the component is less than the magnitude of the total angular momentum. Therefore, the angular momentum vector can never coincide with the axis with respect to which the observations are made. Thus the x and y components of the angular momentum are not zero. This is illustrated in figure which shows how the angular momentum vector may be oriented with respect to the z-axis for the case m = l = 2. When the atom is in a magnetic field, the field exerts a torque on the magnetic dipole of the atom. This torque causes the magnetic dipole and hence the angular momentum vector to precess or rotate about the direction of the magnetic field. This effect is analogous to the precession of a child's top which is spinning with its axis (and hence its angular momentum vector) at an angle to the earth's gravitational field. In this case the gravitational field exerts the torque and the axis of the top slowly revolves around the perpendicular direction as indicated in the figure. The angle of inclination of M with respect to the field direction remains constant during the precession. The z-component of M is therefore constant but the x and y components are continuously changing. Because of the precession, only one component of the electronic angular momentum of an atom can be determined in a given experiment.

The quantum number m determines the magnitude of the component of the angular momentum along a given axis in space. Therefore, it is not surprising that this same quantum number determines the axis along which the electron density is concentrated. When $m = 0$ for a p electron (regardless of the n value, $2p$, $3p$, $4p$, etc). the electron density distribution is concentrated along the z-axis implying that the classical axis of rotation must lie in the x-y plane. Thus a p electron with $m = 0$ is most likely to be found along one axis and has a zero probability of being on the remaining two axes. The effect of the angular momentum possessed by the electron is to concentrate density along one axis. When $m = 1$ or -1 the density distribution of a p electron is concentrated in the x-y plane with doughnut-shaped circular contours. The $m = 1$ and -1 density distributions are identical in appearance. Classically they differ only in the direction of rotation of the electron around the z-axis; counter-clockwise for $m = +1$ and clockwise for $m = -1$. This explains why they have magnetic moments with their north poles in opposite directions.

We can obtain density diagrams for the $m = +1$ and -1 cases similar to the $m = 0$ case by removing the resultant angular momentum component along the z-axis. We can take combinations of the $m = +1$ and -1 functions such that one combination is concentrated along the x-axis and the other along the y-axis, and both are identical to the $m = 0$ function in their appearance. Thus these functions are often labelled as p_x, p_y and p_z functions rather than by their m values. The m value is, however, the true quantum number and we are cheating physically by labelling them p_x, p_y and p_z. This would correspond to applying the field first in the z direction, then in the x direction and finally in the y direction and trying to save up the information each time. In reality when the direction of the field is changed, all the information regarding the previous direction is lost and every atom will again align itself with one chance out of three of being in one of the possible component states with respect to the new direction.

We should note that the r dependence of the orbitals changes with changes in n or l, but the directional component changes with l and m only. Thus all s orbitals possess spherical charge distributions and all p orbitals possess dumb-bell shaped charge distributions regardless of the value of n.

Table: The atomic orbitals for the hydrogen atom.

E_n	n	l	m		Symbol for orbital	
-K	1	0	0		1s	

	n	l	m		orbital	
	2	0	0		2s	
$-\dfrac{1}{4}K$	2	1	1		$2p_{+1}$	$\Big\}\, p_x, p_y, p_z$
	2	1	0		$2p_0$	
	2	1	-1		$2p_{-1}$	
	3	0	0		3s	
	3	1	1		$3p_{+1}$	$\Big\}\, p_x, p_y, p_z$
	3	1	0		$3p_0$	
$-\dfrac{1}{9}K$	3	1	-1		$3p_{-1}$	
	3	2	2		$3d_{+2}$	$\Big\}\, d_{z^2}, d_{x^2-y^2}2, d_{xy}, d_{xz}, d_{yz}$
	3	2	1		$3d_{+1}$	
	3	2	0		$3d_0$	
	3	2	-1		$3d_{-1}$	
	3	2	-2		$3d_{-2}$	

Table summarizes the allowed combinations of quantum numbers for an electron in a hydrogen atom for the first few values of n; the corresponding name (symbol) is given for each orbital. Notice that there are n^2 orbitals for each value of n, all of which belong to the same quantum level and have the same energy. There are $n - 1$ values of l for each value of n and there are $(2l + 1)$ values of m for each value of l. Notice also that for every increase in the value of n, orbitals of the same l value (same directional dependence) as found for the preceding value of n are repeated. In addition, a new value of l and a new shape are introduced. Thus there is a repetition in the shapes of the density distributions along with an increase in their number. We can see evidence of a periodicity in these functions (a periodic re-occurrence of a given density distribution) which we might hope to relate to the periodicity observed in the chemical and physical properties of the elements. We might store this idea in the back of our minds until later.

We can summarize what we have found so far regarding the energy and distribution of an electron in a hydrogen atom thus:

- The energy increases as n increases, and depends only on n, the principal quantum number.

- The average value of the distance between the electron and the nucleus increases as n increases.

- The number of nodes in the probability distribution increases as n increases.

- The electron density becomes concentrated along certain lines (or in planes) as l is increased.

Some words of caution about energies and angular momentum should be added. In passing from the domain of classical mechanics to that of quantum mechanics we retain as many of the familiar words as possible. Examples are kinetic and potential energies, momentum, and angular momentum. We must, however, be on guard when we use these familiar concepts in the atomic domain. All have an altered meaning. Let us make this clear by considering these concepts for the hydrogen atom.

Perhaps the most surprising point about the quantum mechanical expression for the energy is that it does not involve r, the distance between the nucleus and the electron. If the system were a classical one, then we would expect to be able to write the total energy E_n as:

$$E_n, KE = PE = -\frac{1}{2}mv^2 - \frac{e^2}{r}$$

Both the KE and PE would be functions of r, i.e., both would change in value as r was changed (corresponding to the motion of the electron). Furthermore, the sum of the PE and KE must always yield the same value of E_n which is to remain constant.

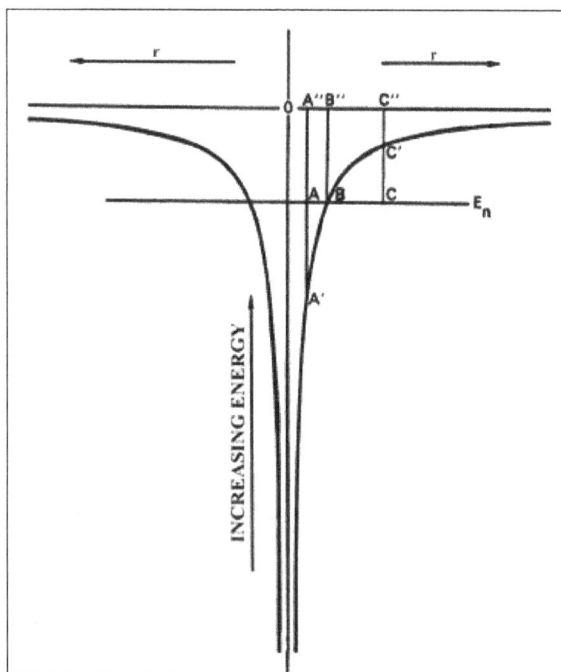

The potential energy diagram for an H atom with one of the allowed energy values superimposed on it.

Fig is the potential energy diagram for the hydrogen atom and we have superimposed on it one of the possible energy levels for the atom, E_n. Consider a classical value for r at the point A". Classically, when the electron is at the point A", its PE is given by the value of the PE curve at A'. The KE is thus equal to the length of the line $A - A'$ in energy units. Thus the sum of $PE + KE$ adds up to E_n.

When the electron is at the point B", its PE would equal E_n and its KE would be zero. The electron would be motionless. Classically, for this value of E_n the electron could not increase its value of r beyond the point represented by B". If it did, it would be inside the "potential wall". For example, consider the point C". At this value of r, the PE is given by the value at C' which is now greater than E_n and hence the KE must be equal to the length of the line $C - C'$. But the KE must now be negative in sign so that the sum of PE and KE will still add up to E_n. What does a negative KE mean? It doesn't mean anything as it never occurs in a classical system. Nor does it occur in a quantum mechanical system. It is true that quantum mechanics does predict a finite probability for the electron being inside the potential curve and indeed for all values of r out to infinity. However, the quantum mechanical expression for E_n does not allow us to determine the instantaneous values for the PE and KE. Instead, we can determine only their average values. Thus quantum mechanics does not give equation (E_n, $KE = PE = -\dfrac{1}{2}mv^2 - \dfrac{e^2}{r}$) but instead states only that the average potential and kinetic energies may be known:

$$E_n = \overline{PE} = \overline{KE}\ \overline{r}$$

A bar denotes the fact that the energy quantity has been averaged over the complete motion (all values of r) of the electron.

Why can r not appear in the quantum mechanical expression for E_n, and why can we obtain only average values for the KE and PE? When the electron is in a given energy level its energy is precisely known; it is E_n. The uncertainty in the value of the momentum of the electron is thus at a minimum. Under these conditions we have seen that our knowledge of the position of the electron is very uncertain and for an electron in a given energy level we can say no more about its position than that it is bound to the atom. Thus if the energy is to remain fixed and known with certainty, we cannot, because of the uncertainty principle, refer to (or measure) the electron as being at some particular distance r from the nucleus with some instantaneous values for its PE and KE. Instead, we may have knowledge of these quantities only when they are averaged over all possible positions of the electron. This discussion again illustrates the pitfalls (e.g., a negative kinetic energy) which arise when a classical picture of an electron as a particle with a definite instantaneous position is taken literally.

It is important to point out that the classical expressions which we write for the dependence of the potential energy on distance, $-e^2/r$ for the hydrogen atom for example, are the expressions employed in the quantum mechanical calculation. However, only the average value of the PE may be calculated and this is done by calculating the value of $-e^2/r$ at every point in space, taking into account the fraction of the total electronic charge at each point in space. The amount of charge at a given point in three-dimensional space is, of course, determined by the electron density distribution. Thus the value of $\overline{PE}$ for the ground state of the hydrogen atom is the electrostatic energy of interaction between a nucleus of charge $+1e$ with the surrounding spherical distribution of negative charge.

We can say more about the $\overline{PE}$ and $\overline{KE}$ for an electron in an atom. Not only are these values constant for a given value of n, but also for any value of n.

$$\overline{KE} = \frac{1}{2}\overline{PE} = -En$$

Thus the $\overline{KE}$ is always positive and equal to minus one half of the $\overline{PE}$. Since the total energy E_n is negative when the electron is bound to the atom, we can interpret the stability of atoms as being due to the decrease in the $\overline{PE}$ when the electron is attracted by the nucleus.

The question now arises as to why the electron doesn't "fall all the way" and sit right on the nucleus. When $r = 0$, the $\overline{PE}$ would be equal to minus infinity, and the $\overline{KE}$, which is positive and thus destabilizing, would be zero. Classically this would certainly be the situation of lowest energy and thus the most stable one. The reason for the electron not collapsing onto the nucleus is a quantum mechanical one. If the electron was bound directly to the nucleus with no kinetic energy, its position and momentum would be known with certainty. This would violate Heisenberg's uncertainty principle. The uncertainty principle always operates through the kinetic energy causing it to become large and positive as the electron is confined to a smaller region of space. (Recall that in the example of an electron moving on a line, the $\overline{KE}$ increased as the length of the line decreased). The smaller the region to which the electron is confined, the smaller is the uncertainty in its position. There must be a corresponding increase in the uncertainty of its momentum. This is brought about by the increase in the kinetic energy which increases the magnitude of the momentum and thus the uncertainty in its value. In other words the bound electron must always possess kinetic energy as a consequence of quantum mechanics.

The $\overline{KE}$ and $\overline{PE}$ have opposite dependences on $\overline{r}$. The $\overline{PE}$ decreases (becomes more negative) as $\overline{r}$ decreases but the $\overline{KE}$ increases (making the atom less stable) as $\overline{r}$ decreases. A compromise is reached to make the energy as negative as possible (the atom as stable as possible) and the compromise always occurs when $\overline{KE} = -\frac{1}{2}\overline{PE}$. A further decrease in $\overline{r}$ would decrease the $\overline{PE}$ but only at the expense of a larger increase in the $\overline{KE}$. The reverse is true for an increase in $\overline{r}$. Thus the reason the electron doesn't fall onto the nucleus may be summed up by stating that "the electron obeys quantum mechanics, and not classical mechanics".

References

- A-Many-Electron-Atoms, Map%3A-Chemistry-The-Central-Science-(Brown-et-al.)/06.-Electronic-Structure-of-Atoms, General-Chemistry, Bookshelves: chem.libretexts.org, Retrieved 9 April, 2019

- Shaviv, Glora. The Life of Stars: The Controversial Inception and Emergence of the Theory of Stellar Structure (2010 ed.). Springer. ISBN 978-3642020872

- Halpern, Paul (2017-11-21). "Spin: The Quantum Property That Should Have Been Impossible". Forbes. Starts With A Bang. Retrieved 2018-03-10

- Atomic-Structure, Inorganic, Chemistry: everyscience.com, Retrieved 10 May, 2019

- Andrew, A. V. (2006). "2. Schrödinger equation". Atomic spectroscopy. Introduction of theory to Hyperfine Structure. P. 274. ISBN 978-0-387-25573-6

3

Lone Pair of Electrons

Lone pair of electrons is the pair of valence electrons that are not shared with another atom in a covalent bond. A single lone pair can be found in the nitrogen group, two lone pairs can be found in the chalcogen group and the halogens can carry three lone pairs. This chapter has been carefully written to provide an easy understanding of the related aspects of lone pair of electrons.

A lone pair is an electron pair in the outermost shell of an atom that is not shared or bonded to another atom. It is also called a non-bonding pair. One way to identify a lone pair is to draw a Lewis structure. The number of lone pair electrons added to the number of bonding electrons equals the number of valence electrons of an atom. The lone pair concept is important to valence shell electron pair repulsion (VSEPR) theory, as it helps to explain the geometry of molecules.

Lone Electron Pairs

The central atom also contains one or more pairs of nonbonding electrons, these additional regions of negative charge will behave much like those associated with the bonded atoms. The orbitals containing the various bonding and nonbonding pairs in the valence shell will extend out from the central atom in directions that minimize their mutual repulsions.

AXE methodLone pairs change a molecule's shape.

Coordination Number and the Central Atom

Coordination number refers to the number of electron pairs that surround a given atom, often referred to as the central atom. The geometries of molecules with lone pairs will differ from those without lone pairs, because the lone pair looks like empty space in a molecule. Both classes of geometry are named after the shapes of the imaginary geometric figures (mostly regular solid polygons) that would be centered on the central atom and have an electron pair at each vertex.

In the water molecule (AX_2E_2), the central atom is O, and the Lewis electron dot formula predicts that there will be two pairs of nonbonding electrons. The oxygen atom will therefore be tetrahedrally coordinated, meaning that it sits at the center of the tetrahedron. Two of the coordination positions are occupied by the shared electron-pairs that constitute the O–H bonds, and the other two by the non-bonding pairs. Therefore, although the oxygen atom is tetrahedrally coordinated, the bonding geometry (shape) of the H_2O molecule is described as bent.

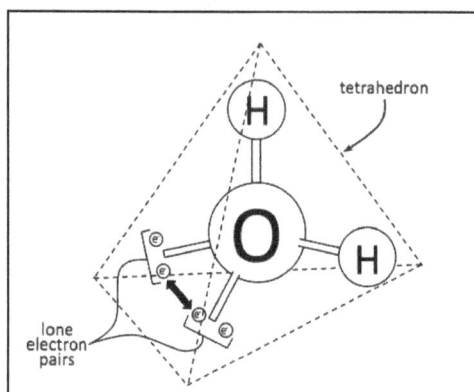

The effect of the lone pair on water, although the oxygen atom is tetrahedrally coordinated, the bonding geometry (shape) of the H_2O molecule is described as bent.

The Repulsive Effect of the Lone Pair Electrons

There is an important difference between bonding and non-bonding electron orbitals. Because a nonbonding orbital has no atomic nucleus at its far end to draw the electron cloud toward it, the charge in such an orbital will be concentrated closer to the central atom; as a consequence, non-bonding orbitals exert more repulsion on other orbitals than do bonding orbitals. In H_2O, the two nonbonding orbitals push the bonding orbitals closer together, making the H–O–H angle 104.5° instead of the tetrahedral angle of 109.5°.

The electron-dot structure of NH_3 places one pair of nonbonding electrons in the valence shell of the nitrogen atom. This means that there are three bonded atoms and one lone pair for a coordination number of four around the nitrogen, the same as occurs in H_2O.

The Lewis dot structure for ammonia, NH3. The lone pair attached to the central nitrogen creates bond angles that differ from the tetrahedral 109.5°.

We can therefore predict that the three hydrogen atoms will lie at the corners of a tetrahedron centered on the nitrogen atom. The lone pair orbital will point toward the fourth corner of the tetrahedron, but since that position will be vacant, the NH_3 molecule itself cannot be tetrahedral; instead, it assumes a pyramidal shape, more specifically, that of a trigonal pyramid (a pyramid with a triangular base). The hydrogen atoms are all in the same plane, with the nitrogen outside of the plane. The non-bonding electrons push the bonding orbitals together slightly, making the H–N–H bond angles about 107°.

In 5-coordinated molecules containing lone pairs, these non-bonding orbitals (which are closer to the central atom and thus more likely to be repelled by other orbitals) will preferentially reside in the equatorial plane. This will place them at 90° angles with respect to no more than two axially-oriented bonding orbitals. We can therefore predict that an AX4E molecule (one in which the central atom A is coordinated to four other atoms X and to one nonbonding electron pair) such as SF4 will have a "see-saw" shape.

Ammonium Ion

The ammonium cation is a positively charged polyatomic ion with the chemical formula NH_4^+. It is formed by the protonation of ammonia (NH_3). Ammonium is also a general name for positively charged or protonated substituted amines and quaternary ammonium cations (NR_4^+), where one or more hydrogen atoms are replaced by organic groups (indicated by R).

Acid–base Properties

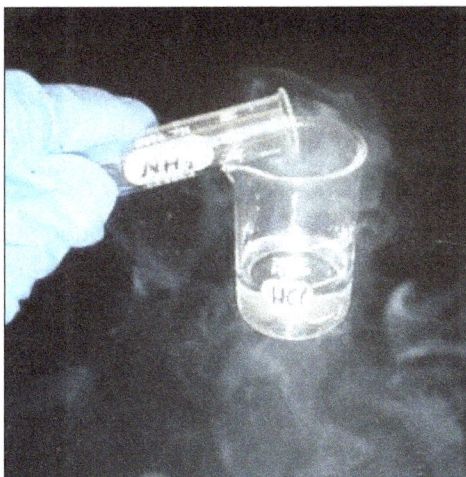

Fumes from hydrochloric acid and ammonia forming a white cloud of ammonium chloride.

The ammonium ion is generated when ammonia, a weak base, reacts with Brønsted acids (proton donors):

$$H^+ + NH_3 \rightarrow NH_4^+$$

The ammonium ion is mildly acidic, reacting with Brønsted bases to return to the uncharged ammonia molecule:

$$NH_4^+ + B^- \rightarrow HB + NH_3$$

Thus, treatment of concentrated solutions of ammonium salts with strong base gives ammonia. When ammonia is dissolved in water, a tiny amount of it converts to ammonium ions:

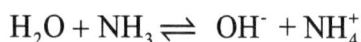

$$H_2O + NH_3 \rightleftharpoons OH^- + NH_4^+$$

The degree to which ammonia forms the ammonium ion depends on the pH of the solution. If the pH is low, the equilibrium shifts to the right: more ammonia molecules are converted into ammonium ions. If the pH is high (the concentration of hydrogen ions is low), the equilibrium shifts to the left: the hydroxide ion abstracts a proton from the ammonium ion, generating ammonia.

Formation of ammonium compounds can also occur in the vapor phase; for example, when ammonia vapor comes in contact with hydrogen chloride vapor, a white cloud of ammonium chloride forms, which eventually settles out as a solid in a thin white layer on surfaces.

Ammonium Salts

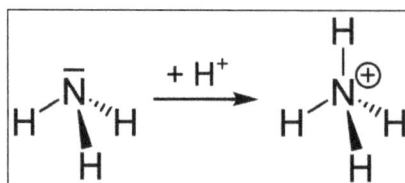

Formation of ammonium.

Ammonium cation is found in a variety of salts such as ammonium carbonate, ammonium chloride and ammonium nitrate. Most simple ammonium salts are very soluble in water. An exception is ammonium hexachloroplatinate, the formation of which was once used as a test for ammonium. The ammonium salts of nitrate and especially perchlorate are highly explosive, in these cases ammonium is the reducing agent.

In an unusual process, ammonium ions form an amalgam. Such species are prepared by the electrolysis of an ammonium solution using a mercury cathode. This amalgam eventually decomposes to release ammonia and hydrogen.

Structure and Bonding

The lone electron pair on the nitrogen atom (N) in ammonia, represented as a line above the N, forms the bond with a proton (H$^+$). Thereafter, all four N–H bonds are equivalent, being polar covalent bonds. The ion has a tetrahedral structure and is isoelectronic with methane and borohydride. In terms of size, the ammonium cation (r_{ionic} = 175 pm) resembles the caesium cation (r_{ionic} = 183 pm).

Organic Ammonium Ions

The hydrogen atoms in the ammonium ion can be substituted with an alkyl group or some other organic group to form a substituted ammonium ion (IUPAC nomenclature: aminium ion). Depending on the number of organic groups, the ammonium cation is called a primary, secondary, tertiary, or quaternary. With the exception of the quaternary ammonium cations, the organic ammonium cations are weak acids.

An example of a reaction forming an ammonium ion is that between dimethylamine, $(CH_3)_2NH$, and an acid to give the dimethylaminium cation, $(CH_3)_2 NH_2^+$:

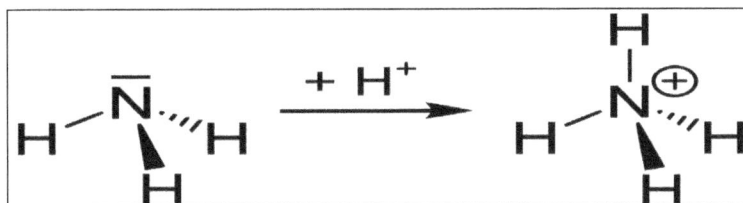

Quaternary ammonium cations have four organic groups attached to the nitrogen atom, they lack a hydrogen atom bonded to the nitrogen atom. These cations, such as the tetra-*n*-butylammonium cation, are sometimes used to replace sodium or potassium ions to increase the solubility of the associated anion in organic solvents. Primary, secondary, and tertiary ammonium salts serve the same function, but are less lipophilic. They are also used as phase-transfer catalysts and surfactants.

An unusual class of organic ammonium salts are derivatives of amine radical cations, $R_3N^{+\cdot}$ such as tris(4-bromophenyl)ammonium hexachloroantimonate.

Biology

Ammonium ions are a waste product of the metabolism of animals. In fish and aquatic invertebrates, it is excreted directly into the water. In mammals, sharks, and amphibians, it is converted in the urea cycle to urea, because urea is less toxic and can be stored more efficiently. In birds, reptiles, and terrestrial snails, metabolic ammonium is converted into uric acid, which is solid and can therefore be excreted with minimal water loss.

Ammonium is an important source of nitrogen for many plant species, especially those growing on hypoxic soils. However, it is also toxic to most crop species and is rarely applied as a sole nitrogen source.

Ammonium Metal

The ammonium ion has very similar properties to the heavier alkali metals and is often considered a close relative. Ammonium is expected to behave as a metal (NH_4^+ ions in a sea of electrons) at very high pressures, such as inside gas giant planets such as Uranus and Neptune.

Under normal conditions, ammonium does not exist as a pure metal, but does as an amalgam (alloy with mercury).

Water Molecule

Water is a chemical compound and polar molecule, which is liquid at standard temperature and pressure. It has the chemical formula H_2O, meaning that one molecule of water is composed of two hydrogen atoms and one oxygen atom. Water is found almost everywhere on earth and is required by all known life. About 70% of the Earth's surface is covered by water. Water is known to exist, in ice form, on several other bodies in the solar system and beyond, and proof that it exists (or did exist) in liquid form anywhere besides Earth would be strong evidence of extraterrestrial life.

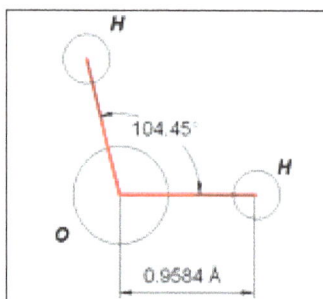

The solid state of water is known as ice; the gaseous state is known as water vapor (or steam). The units of temperature (formerly the degree Celsius and now the Kelvin) are defined in terms of the triple point of water, 273.16 K (0.01 °C) and 611.2 Pa, the temperature and pressure at which solid, liquid, and gaseous water coexist in equilibrium. Water exhibits some very strange behaviors, including the formation of states such as vitreous ice, a noncrystalline (glassy), solid state of water.

At temperatures greater than 647 K and pressures greater than 22.064 MPa, a collection of water molecules assumes a supercritical condition, in which liquid-like clusters float within a vapor-like phase.

The liquid water path is a measure of the amount of liquid water in an air column.

The Dipolar Nature of the Water Molecule

An important feature of the water molecule is its polar nature. The water molecule forms an angle, with hydrogen atoms at the tips and oxygen at the vertex. Since oxygen has a higher electronegativity than hydrogen, the side of the molecule with the oxygen atom has a partial negative charge. A molecule with such a charge difference is called a dipole. The charge differences cause water molecules to be attracted to each other (the relatively positive areas being attracted to the relatively negative areas) and to other polar molecules. This attraction is known as hydrogen bonding.

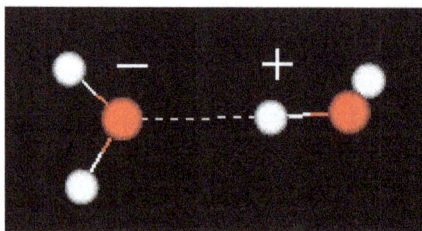

hydrogen bond between two water molecules.

This relatively weak (relative to the covalent bonds within the water molecule itself) attraction results in physical properties such as a relatively high boiling point, because a lot of heat energy is necessary to break the hydrogen bonds between molecules. For example, sulfur is the element below oxygen in the periodic table, and its equivalent compound, hydrogen sulfide (H_2S) does not have hydrogen bonds, and though it has twice the molecular weight of water, it is a gas at room temperature. The extra bonding between water molecules also gives liquid water a large specific heat capacity.

Hydrogen bonding also gives water molecules an unusual behaviour when freezing. Just like most other materials, the liquid becomes denser with lowering temperature. However, unlike most other materials, when cooled to near freezing point, the presence of hydrogen bonds means that the molecules, as they rearrange to minimise their energy, form a structure that is actually of lower density: hence the solid form, ice, will float in water. In other words, water expands as it freezes (most other materials shrink on solidification). Liquid water reaches its highest density at a temperature of 4 °C. This has an interesting consequence for water life in winter. Water chilled at the surface becomes denser and sinks, forming convection currents that cool the whole water body, but when the temperature of the lake water reaches 4°C, water on the surface, as it chills further, becomes less dense, and stays as a surface layer which eventually forms ice. Since downward convection of colder water is blocked by the density change, any large body of water frozen in winter will have the bulk of its water still liquid at 4°C beneath the icy surface, allowing fish to survive. This is one of the principal examples of finely-tuned physical properties that support life on Earth that is used as an argument for the anthropic principle.

Another consequence is that ice will melt if sufficient pressure is applied.

Structure of Water and Ice

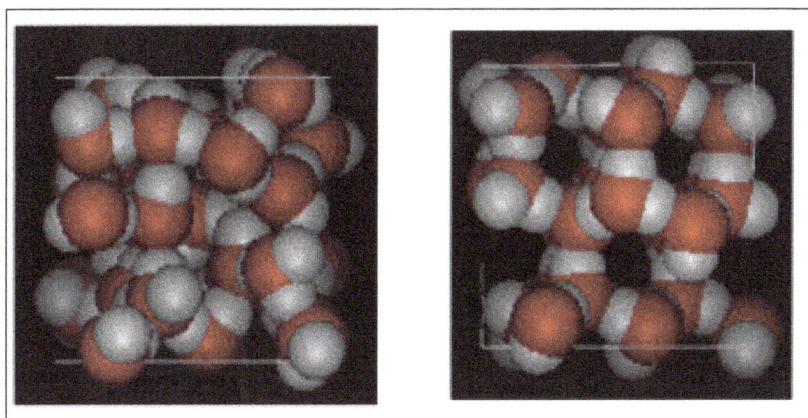

Shown above is a side by side comparison of a box 10 Angstroms across. It clearly shows that ice takes up more space because of the hydrogen bonding that occurs when the state changes from liquid to solid. In ice Ih, each water forms four hydrogen bonds with O---O distances of 2.76 Angstroms to the nearest oxygen neighbor. Because of ordered structure in ice there are less H_2O molecules in a given space of volume.

Water as a Solvent

Water is also a good solvent due to its polarity. The solvent properties of water are vital in biology, because many biochemical reactions take place only within aqueous solutions (e.g., reactions in the cytoplasm and blood). In addition, water is used to transport biological molecules.

When an ionic or polar compound enters water, it is surrounded by water molecules. The relatively small size of water molecules typically allows many water molecules to surround one molecule of solute. The partially negative dipoles of the water are attracted to positively charged components of the solute, and vice versa for the positive dipoles.

In general, ionic and polar substances such as acids, alcohols, and salts are easily soluble in water, and nonpolar substances such as fats and oils are not. Nonpolar molecules stay together in water because it is energetically more favorable for the water molecules to hydrogen bond to each other than to engage in van der Waals interactions with nonpolar molecules.

An example of an ionic solute is table salt; the sodium chloride, NaCl, separates into Na+ cations and Cl- anions, each being surrounded by water molecules. The ions are then easily transported away from their crystalline lattice into solution. An example of a nonionic solute is table sugar. The water dipoles hydrogen bond to the dipolar regions of the sugar molecule and allow it to be carried away into solution.

Cohesion and Surface Tension

The strong hydrogen bonds give water a high cohesiveness and, consequently, surface tension. This is evident when small quantities of water are put onto a nonsoluble surface and the water stays together as drops. This feature is important when water is carried through xylem up stems in plants; the strong intermolecular attractions hold the water column together, and prevent tension caused by transpiration pull. Other liquids with lower surface tension would have a higher tendency to "rip", forming vacuum or air pockets and rendering the xylem vessel inoperative.

Conductivity

Pure water is actually a good insulator (poor conductor), meaning that it does not conduct electricity well. Because water is such a good solvent, however, it often has some solute dissolved in it, most frequently salt. If water has such impurities, then it can conduct electricity much better, because impurities such as salt comprise free ions in aqueous solution by which an electric current can flow.

Electrolysis

Water can be split into its constituent elements, hydrogen and oxygen, by passing a current through it. This process is called electrolysis. Water molecules naturally disassociate into H+ and OH- ions, which are pulled toward the cathode and anode, respectively. At the cathode, two H+ ions pick up electrons and form H_2 gas. At the anode, four OH- ions combine and release O_2 gas, molecular water, and four electrons. The gases produced bubble to the surface, where they can be collected.

Reactivity

Chemically, water is amphoteric: able to act as an acid or base. Occasionally the term hydroxic acid is used when water acts as an acid in a chemical reaction. At a pH of 7 (neutral), the concentration of hydroxide ions (OH-) is equal to that of the hydronium (H_3O+) or hydrogen ions (H+) ions. If the equilibrium is disturbed, the solution becomes acidic (higher concentration of hydronium ions) or basic (higher concentration of hydroxide ions).

Water can act as either an acid or a base in reactions. According to the Brønsted-Lowry system, an acid is defined as a species which donates a proton (an H+ ion) in a reaction, and a base as one which receives a proton. When reacting with a stronger acid, water acts as a base; when reacting with a weaker acid, it acts as an acid. For instance, it receives an H+ ion from HCl in the equilibrium:

$$HCl + H_2O \rightarrow H_3O^+ + Cl^-$$

Here water is acting as a base, by receiving an H+ ion. An acid donates an H+ ion, and water can also do this, such as in the reaction with ammonia, NH3:

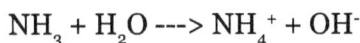

$$NH_3 + H_2O ---> NH_4^+ + OH^-$$

pH in Practice

In theory, pure water has a pH of 7. In practice, pure water is very difficult to produce. Water left exposed to air for any length of time will rapidly dissolve carbon dioxide, forming a solution of carbonic acid, with a limiting pH of ~5.7.

Purifying Water

Purified water is needed for many industrial applications, as well as for consumption. Humans require water that does not contain too much salt or other impurities. Common impurities include chemicals or harmful bacteria. Some solutes are acceptable and even desirable for perceived taste enhancement. Water that is suitable for drinking is termed potable water.

Six popular methods for purifying water are:

- Filtering: Water is passed through a sieve that catches small particles. The tighter the mesh of the sieve, the smaller the particles must be to pass through. Filtering is not sufficient to completely purify water, but it is often a necessary first step, since such particles can interfere with the more thorough purification methods.

- Boiling: Water is heated to its boiling point long enough to inactivate or kill microorganisms that normally live in water at room temperature. In areas where the water is "hard", (containing dissolved calcium salts), boiling decomposes the bicarbonate ion, resulting in some (but not all) of the dissolved calcium being precipitated in the form of calcium carbonate. This is the so-called "fur" that builds up on kettle elements etc. in hard water areas. With the exception of calcium, boiling does not remove solutes of higher boiling point than water, and in fact increases their concentration (due to some water being lost as vapour)

- Carbon filtering: Charcoal, a form of carbon with a high surface area due to its mode of preparation, adsorbs many compounds, including some toxic compounds. Water is passed through activated charcoal to remove such contaminants. This method is most commonly used in household water filters and fish tanks. Household filters for drinking water sometimes also contain silver, trace amounts of silver ions having a bactericidal effect.

- Distilling: Distillation involves boiling the water to produce water vapour. The water vapour then rises to a cooled surface where it can condense back into a liquid and be collected. Because the solutes are not normally vaporized, they remain in the boiling solution.

Even distillation does not completely purify water, because of contaminants with similar boiling points and droplets of unvaporized liquid carried with the steam. However, 99.9% pure water can be obtained by distillation.

- Reverse osmosis: Mechanical pressure is applied to an impure solution to force pure water through a semi-permeable membrane. The term is reverse osmosis, because normal osmosis would result in pure water moving in the other direction to dilute the impurities. Reverse osmosis is theoretically the most thorough method of large-scale water purification available, although perfect semi-permable membranes are difficult to create. on exchange chromatography: In this case, water is passed through a charged resin column that has side chains that trap calcium, magnesium, and other heavy metal ions. In many laboratories, this method of purification has replaced distillation, as it provides a high volume of very pure water more quickly and with less energy use than other processes. Water purified in this way is called deionized water.

Wasting Water

Wasting water is the abuse of water, i.e. using it unnecessarily. An example is the use of water, particularly water purified to human safe drinking standards, in unnecessary irrigation. Also, in homes, water may be wasted if the toilet is flushed unnecessarily or the tank leaks. Causing water to become polluted may be the biggest single abuse of water. To the extent that a pollutant limits other uses of the water, it becomes a waste of the resource, regardless of benefits to the polluter.

Lone Pair on Central Atom

The molecular geometries of molecules change when the central atom has one or more lone pairs of electrons. The total number of electron pairs, both bonding pairs and lone pairs, leads to what is called the electron domain geometry. When one or more of the bonding pairs of electrons is replaced with a lone pair, the molecular geometry (actual shape) of the molecule is altered. A subscript will be used when there is more than one lone pair. Lone pairs on the surrounding atoms (B) do not affect the geometry.

AB_3E: Ammonia, NH_3

The ammonia molecule contains three single bonds and one lone pair on the central nitrogen atom.

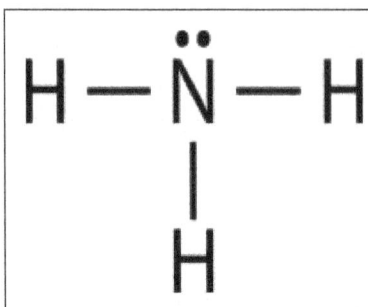

Lone pair electrons in ammonia.

The domain geometry for a molecule with four electron pairs is tetrahedral, as was seen with CH_4. In the ammonia molecule, one of the electron pairs is a lone pair rather than a bonding pair. The molecular geometry of NH_3 is called trigonal pyramidal.

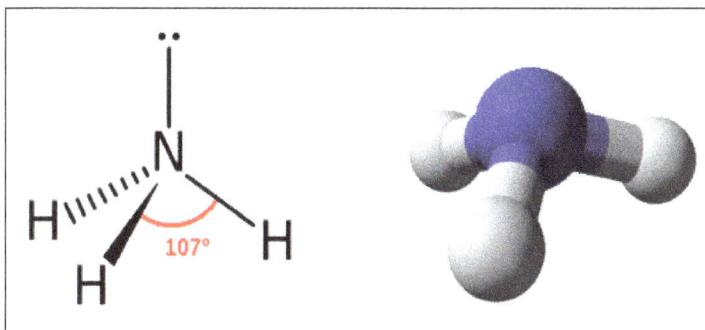
Ammonia molecule.

Recall that the bond angle in the tetrahedral CH_4 molecule is 109.5°. Again, the replacement of one of the bonded electron pairs with a lone pair compresses the angle slightly. The H-N-H angle is approximately 107°.

AB_2E_2: Water, H_2O

A water molecule consists of two bonding pairs and two lone pairs.

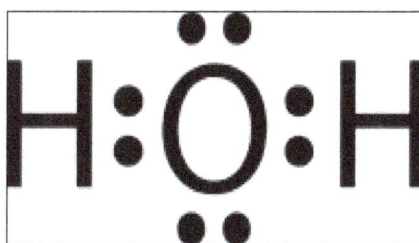
Lone pair electrons on water.

As for methane and ammonia, the domain geometry for a molecule with four electron pairs is tetrahedral. In the water molecule, two of the electron pairs are lone pairs rather than bonding pairs. The molecular geometry of the water molecule is bent. The H-O-H bond angle is 104.5°, which is smaller than the bond angle in NH_3.

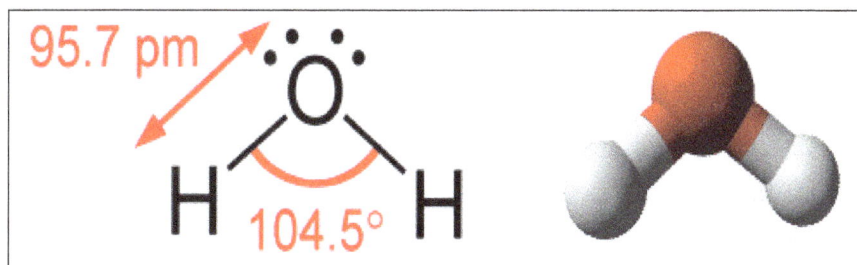
Water molecule.

AB_4E: Sulfur Tetrafluoride, SF_4

The Lewis structure for SF_4 contains four single bonds and a lone pair on the sulfur atom.

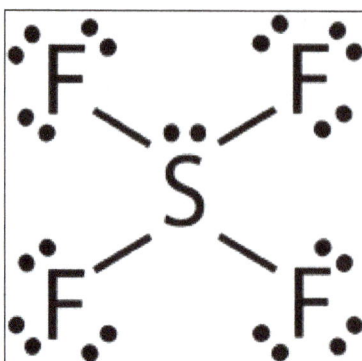
Lone pair electrons in SF_4.

The sulfur atom has five electron groups around it, which corresponds to the trigonal bipyramidal domain geometry, as in PCl_5. Recall that the trigonal bipyramidal geometry has three equatorial atoms and two axial atoms attached to the central atom. Because of the greater repulsion of a lone pair, it is one of the equatorial atoms that are replaced by a lone pair. The geometry of the molecule is called a distorted tetrahedron or seesaw.

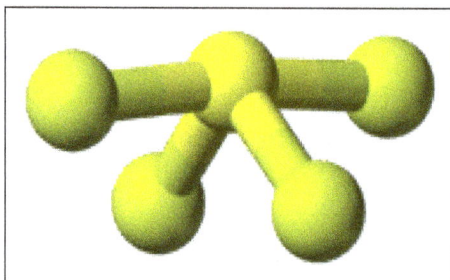
Ball and stick model for SF_4.

Table: Geometries in which the central atom has one or more lone pairs.

Total Number of Electron Pairs	Number of Bonding Pairs	Number of Lone Pairs	Electron Domain Geometry	Molecular Geometry	Examples
3	2	1	trigonal planar	bent	O_3
4	3	1	tetrahedral	trigonal pyramidal	NH_3
4	2	2	tetrahedral	bent	H_2O
5	4	1	trigonal bipyramidal	distorted tetrahedron (seesaw)	SF_4
5	3	2	trigonal bipyramidal	T-shaped	ClF_3
5	2	3	trigonal bipyramidal	linear	I_3^-
6	5	1	octahedral	square pyramidal	BrF_5
6	4	2	octahedral	square planar	XeF_4

References

- Definition-of-lone-pair-605314: thoughtco.com, Retrieved 11 June, 2019

- Holleman, Arnold Frederik; Wiberg, Egon (2001), Wiberg, Nils (ed.), Inorganic Chemistry, translated by Eagleson, Mary; Brewer, William, San Diego/Berlin: Academic Press/De Gruyter, ISBN 0-12-352651-5

- Lone-electron-pairs, chapter, introchem: courses.lumenlearning.com, Retrieved 12 July, 2019

- Britto, DT; Kronzucker, HJ (2002). "NH4+ toxicity in higher plants: a critical review" (PDF). Journal of Plant Physiology. 159 (6): 567–584. doi:10.1078/0176-1617-0774

- Water, solvents: worldofmolecules.com, Retrieved 13 August, 2019

- Molecular-Shapes:-Lone-Pairs-on-Central-Atom-CHEM, lesson, molecular-shapes-lone-pairs-on-central-atom, chemistry: ck12.org, Retrieved 14 January, 2019

4
Types of Bonds

Bond is referred to as the attraction between atoms, ions and molecules to form a chemical compound. It can be categorized into pi bond, sigma bond, single bond, double bond and triple bond. This chapter delves into these types of bonds to provide an in-depth understanding of the subject.

Pi Bond

Pi bonds are chemical bonds that are covalent in nature and involve the lateral overlapping of two lobes of an atomic orbital with two lobes of another atomic orbital that belongs to a different atom. Pi bonds are often written as 'π bonds'.

The two orbitals which are bonded share the same nodal plane at which the electron density is zero. This plane passes through the nuclei of the two bonded atoms and is also the nodal plane for the molecular orbitalcorresponding to the pi bond.

An illustration depicting the overlapping of two different p orbitals in order to form a π bond is given below.

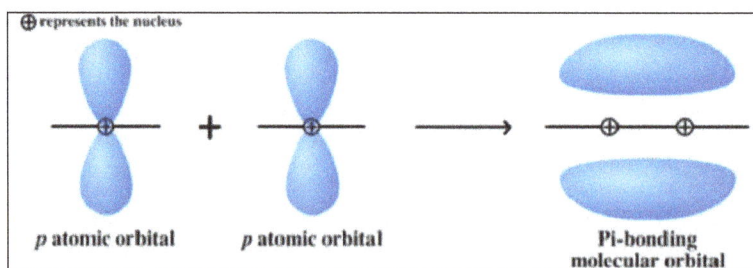

Commonly, pi bonding involves p orbitals. However, d orbitals also have the ability to participate in π bonding and these types of bonds involving d orbitals can be found in the multiple bonds formed between two metals.

Examples of Pi Bonding

Bonding in Ethene (C_2H_4)

Ethene is often considered the simplest alkene since it contains only 2 carbon atoms (that are doubly bonded to each other) and four hydrogen atoms. The electron configuration of carbon is $1s^2 2s^2 2p^2$. However, in its excited state, an electron is promoted from the 2s orbital to the 2p orbital.

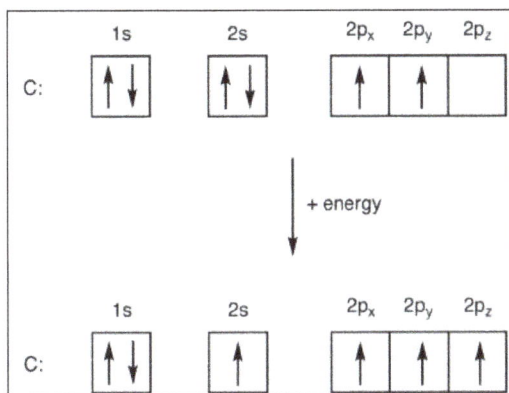

This promotion of a lone pair of 2s electrons occurs when a photon of a specific wavelength transfers energy to the 2s electron, enabling the jump to the 2p orbital. This promotion doesn't require much energy since the energy gap between the 2s and 2p orbital is very small.

Now, the excited carbon atoms undergo sp^2 hybridization to form an sp^2 hybridized molecular orbital (which is made up of a single 's' orbital and two 'p' orbitals). The sp^2 hybridized carbons now form three sigma bonds and one pi bond, as illustrated below.

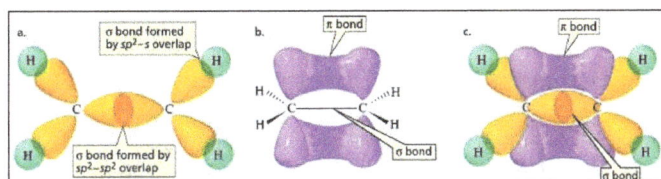

The sp^2 hybridized orbital in the carbon atom is made up of a 2s electron, a $2p_x$ electron, and a $2p_y$ electron. It can form a total of three sigma bonds. The $2p_z$ electrons of the carbon atoms now form a pi bond with each other. Thus, each carbon atom in the ethene molecule participates in three sigma bonds and one pi bond.

Bonding in Ethyne (C_2H_2)

Ethene is the simplest alkyne in which each carbon atom is singly bonded to one hydrogen atom and triply bonded to the other carbon atom. These bonds are formed when photons of a specific wavelength transfer energy to the 2s electron, enabling it to jump to the $2p_z$ orbital.

Now, the 2s orbital and one 2p orbital undergo hybridization to form an sp hybridized orbital. This orbital contains the 2s electron and the $2p_x$ electron. The $2p_y$ and $2p_z$ electrons of the carbon atoms now form pi bonds with each other, as illustrated below.

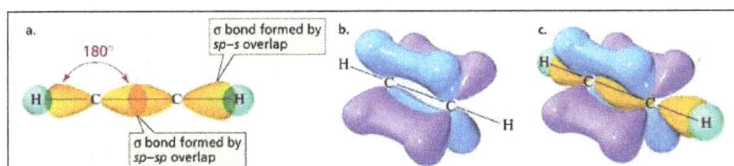

The 'sp' hybridized orbital in the carbon atom can form a total of two sigma bonds. One sigma bond is formed with the adjacent carbon atom and the other is formed with the '1s' orbital belonging to the hydrogen atom. Two pi bonds are formed from the lateral overlap of the $2p_y$ and $2p_z$ orbitals.

Strength of Pi Bonds

The pi bonds are almost always weaker than sigma bonds. For example, two times the bond energy of a carbon-carbon single bond (sigma bond) is greater than a carbon-carbon double bond containing one sigma and one pi bond. This observation on the bond strength of π bonds suggests that they do not add as much stability as sigma bonds do.

The relative weakness of these bonds, when compared to sigma bonds, can be explained with the help of the quantum mechanical perspective that there is a significantly lower degree of overlapping of p orbitals in pi bonds due to the fact that they are oriented parallel to each other. Sigma bonds, however, have a far greater degree of overlapping and hence tend to be stronger than the corresponding π bonds.

Pi Bonding in Multiple Bonds

The following multiple bonds can be broken down into sigma, pi, and delta bonds:

- Generally, double bonds consist of a single sigma bond and a single pi bond. An example of such a bond can be seen in ethylene (or ethene).

- Typically, triple bonds consist of a single sigma bond and two π bonds which are placed in planes that are perpendicular to each other and contain the bond axis.

- Quadruple bonds are very rare bonds which can only be found in the bonds between two transition metal atoms. It can be broken down into one sigma, two pi, and one delta bond.

The combination of pi and sigma bonds in multiple bonds are always stronger than a single sigma bond. The reduction in bond lengths in multiple bonds points towards this statement. This contraction in the bond length can be observed in the length of the carbon-carbon bond in ethane, ethylene, and acetylene, which are equal to 153.51 pm, 133.9 pm, and 120.3 pm respectively.

Therefore, it can be understood that multiple bonds shorten the total bond length and strengthen the overall bond between the two atoms.

Sigma Bond

In chemistry, sigma bonds (σ bonds) are the strongest type of covalent chemical bond. They are formed by head-on overlapping between atomic orbitals. Sigma bonding is most simply defined for diatomic molecules using the language and tools of symmetry groups. In this formal approach, a σ-bond is symmetrical with respect to rotation about the bond axis. By this definition, common forms of sigma bonds are s+s, p_z+p_z, s+p_z and $d_z^2+d_z^2$ (where z is defined as the axis of the bond or the internuclear axis). Quantum theory also indicates that molecular orbitals (MO) of identical symmetry actually mix or hybridize. As a practical consequence of this mixing of diatomic molecules, the wavefunctions s+s and p_z+p_z molecular orbitals become blended. The extent of this mixing (or hybridization or blending) depends on the relative energies of the MOs of like symmetry.

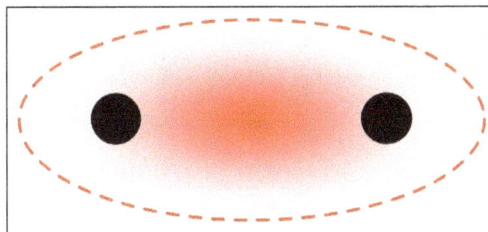

σ bond between two atoms: localization of electron density.

For homodiatomics, bonding σ orbitals have no nodal planes at which the wavefunction is zero, either between the bonded atoms or passing through the bonded atoms. The corresponding antibonding, or σ* orbital, is defined by the presence of one nodal plane between the two bonded atoms.

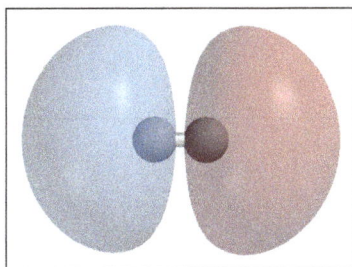

1sσ* antibonding molecular orbital in H_2 with nodal plane.

Sigma bonds are the strongest type of covalent bonds due to the direct overlap of orbitals, and the electrons in these bonds are sometimes referred to as sigma electrons.

The symbol σ is the Greek letter sigma. When viewed down the bond axis, a σ MO has a circular symmetry, hence resembling a similarly sounding "s" atomic orbital.

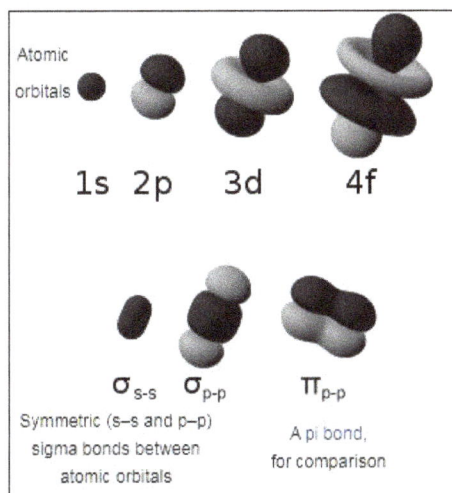

Typically, a single bond is a sigma bond while a multiple bond is composed of one sigma bond together with pi or other bonds. A double bond has one sigma plus one pi bond, and a triple bond has one sigma plus two pi bonds.

$$\sigma_{s-hybrid}$$

$$\sigma_{s-p}$$

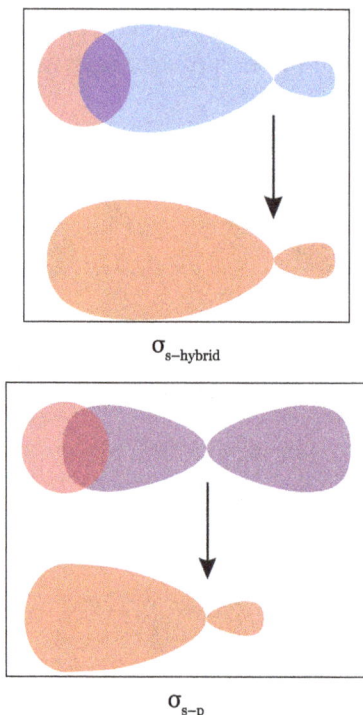

Polyatomic Molecules

Sigma bonds are obtained by head-on overlapping of atomic orbitals. The concept of sigma bonding is extended to describe bonding interactions involving overlap of a single lobe of one orbital with a single lobe of another. For example, propane is described as consisting of ten sigma bonds, one each for the two C–C bonds and one each for the eight C–H bonds.

Multiple-Bonded Complexes

Transition metal complexes that feature multiple bonds, such as the dihydrogen complex, have sigma bonds between the multiple bonded atoms. These sigma bonds can be supplemented with other bonding interactions, such as π-back donation, as in the case of $W(CO)_3(PCy_3)_2(H_2)$, and even δ-bonds, as in the case of chromium(II) acetate.

Organic Molecules

Organic molecules are often cyclic compounds containing one or more rings, such as benzene, and are often made up of many sigma bonds along with pi bonds. According to the **sigma bond rule**, the number of sigma bonds in a molecule is equivalent to the number of atoms plus the number of rings minus one.

$$N_\sigma = N_{atoms} + N_{rings} - 1$$

This rule is a special-case application of the Euler characteristic of the graph which represents the molecule.

A molecule with no rings can be represented as a tree with a number of bonds equal to the number of atoms minus one (as in dihydrogen, H_2, with only one sigma bond, or ammonia, NH_3, with 3 sigma bonds). There is no more than 1 sigma bond between any two atoms.

Molecules with rings have additional sigma bonds, such as benzene rings, which have 6 C–C sigma bonds within the ring for 6 carbon atoms. The anthracene molecule, $C_{14}H_{10}$, has three rings so that the rule gives the number of sigma bonds as $24 + 3 - 1 = 26$. In this case there are 16 C–C sigma bonds and 10 C–H bonds.

This rule fails in the case of molecules which, when drawn flat on paper, have a different number of rings than the molecule actually has - for example, Buckminsterfullerene, C_{60}, which has 32 rings, 60 atoms, and 90 sigma bonds, one for each pair of bonded atoms; however, $60 + 32 - 1 = 91$, not 90. This is because the sigma rule is a special case of the Euler characteristic, where each ring is considered a face, each sigma bond is an edge, and each atom is a vertex. Ordinarily, one extra face is assigned to the space not inside any ring, but when Buckminsterfullerene is drawn flat without any crossings, one of the rings makes up the outer pentagon; the inside of that ring is the outside of the graph. This rule fails further when considering other shapes - toroidal fullerenes will obey the rule that the number of sigma bonds in a molecule is exactly the number of atoms plus the number of rings, as will nanotubes - which, when drawn flat as if looking through one from the end, will have a face in the middle, corresponding to the far end of the nanotube, which is not a ring, and a face corresponding to the outside.

Single Bond

In chemistry, a single bond is a chemical bond between two atoms involving two valence electrons. That is, the atoms share one pair of electrons where the bond forms. Therefore, a single bond is a type of covalent bond. When shared, each of the two electrons involved is no longer in the sole possession of the orbital in which it originated. Rather, both of the two electrons spend time in either of the orbitals which overlap in the bonding process. As a Lewis structure, a single bond is denoted as A:A or A-A, for which A represents an element (Moore, Stanitski, and Jurs 329). In the first rendition, each dot represents a shared electron, and in the second rendition, the bar represents both of the electrons shared in the single bond.

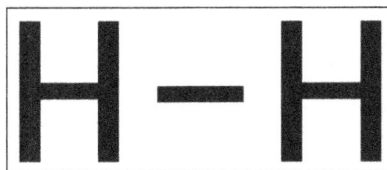

Lewis structure for molecular hydrogen. Note depiction of the single bond.

A covalent bond can also be a double bond or a triple bond. A single bond is weaker than either a double bond or a triple bond. This difference in strength can be explained by examining the component bonds of which each of these types of covalent bonds consists.

Usually, a single bond is a sigma bond. An exception is the bond in diboron, which is a pi bond. In contrast, the double bond consists of one sigma bond and one pi bond, and a triple bond consists of one sigma bond and two pi bonds. The number of component bonds is what determines the strength disparity. It stands to reason that the single bond is the weakest of the three because it consists of only a sigma bond, and the double bond or triple bond consist not only of this type of component bond but also at least one additional bond.

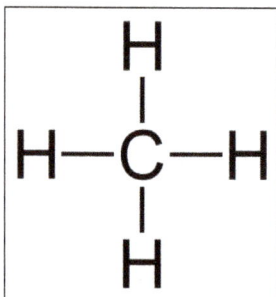

Lewis structure for methane. Note depiction of the four single bonds between the carbon and hydrogen atoms.

The single bond has the capacity for rotation, a property not possessed by the double bond or the triple bond. The structure of pi bonds does not allow for rotation (at least not at 298 K), so the double bond and the triple bond which contain pi bonds are held due to this property. The sigma bond is not so restrictive, and the single bond is able to rotate using the sigma bond as the axis of rotation.

Another property comparison can be made in bond length. Single bonds are the longest of the three types of covalent bonds as interatomic attraction is greater in the two other types, double and triple. The increase in component bonds is the reason for this attraction increase as more electrons are shared between the bonded atoms.

Single bonds are often seen in diatomic molecules. Examples of this use of single bonds include H_2, F_2, and HCl.

Single bonds are also seen in molecules made up of more than two atoms. Examples of this use of single bonds include:

- Both bonds in H_2O.

- All 4 bonds in CH_4.

Single bonding even appears in molecules as complex as hydrocarbons larger than methane. The type of covalent bonding in hydrocarbons is extremely important in the nomenclature of these molecules. Hydrocarbons containing only single bonds are referred to as alkanes (Moore, Stanitski, and Jurs 334). The names of specific molecules which belong to this group end with the suffix –ane' Examples include ethane, 2-methylbutane, and cyclopentane.

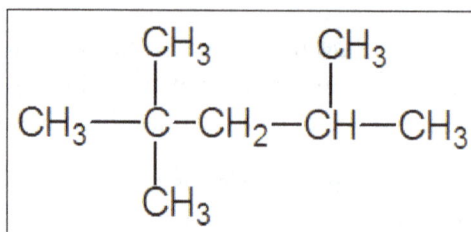

Lewis structure for an alkane. Note that all the bonds are single covalent bonds.

Double Bond

A double bond in chemistry is a chemical bond between two chemical elements involving four bonding electrons instead of the usual two. The most common double bond occurs between two carbon atoms and can be found in alkenes. Many types of double bonds exist between two different elements. For example, in a carbonyl group with a carbon atom and an oxygen atom. Other common double bonds are found in azo compounds (N=N), imines (C=N) and sulfoxides (S=O). In skeletal formula the double bond is drawn as two parallel lines (=) between the two connected atoms; typographically, the equals sign is used for this. Double bonds were first introduced in chemical notation by Russian chemist Alexander Butlerov.

Double bonds involving carbon are stronger than single bonds and are also shorter. The bond order is two. Double bonds are also electron-rich, which makes them potentially more reactive in the presence of a strong electron acceptor (as in addition reactions of the halogens).

Chemical compounds with double bonds

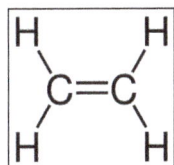 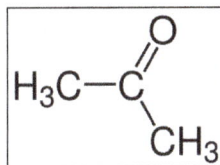 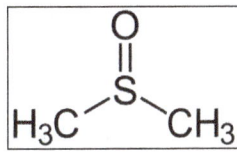 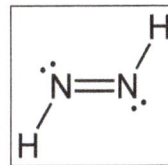

Ethylene	Acetone	Dimethyl sulfoxide	Diazene
Carbon-carbon double bond	Carbon-oxygen double bond	Sulfur-oxygen double bond	Nitrogen-nitrogen double bond

Double Bonds in Alkenes

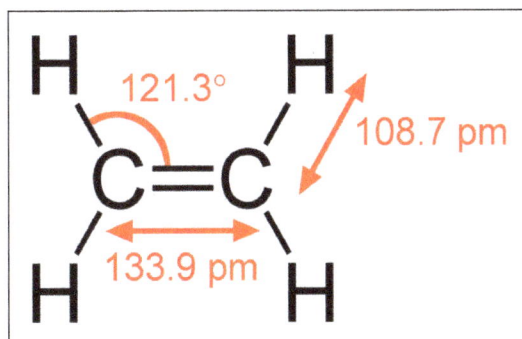

Geometry of ethylene

The type of bonding can be explained in terms of orbital hybridisation. In ethylene each carbon atom has three sp^2 orbitals and one p-orbital. The three sp^2 orbitals lie in a plane with ~120° angles. The p-orbital is perpendicular to this plane. When the carbon atoms approach each other, two of the sp^2 orbitals overlap to form a sigma bond. At the same time, the two p-orbitals approach (again in the same plane) and together they form a pi bond. For maximum overlap, the p-orbitals have to remain parallel, and, therefore, rotation around the central bond is not possible. This property gives rise to cis-trans isomerism. Double bonds are shorter than single bonds because p-orbital overlap is maximized.

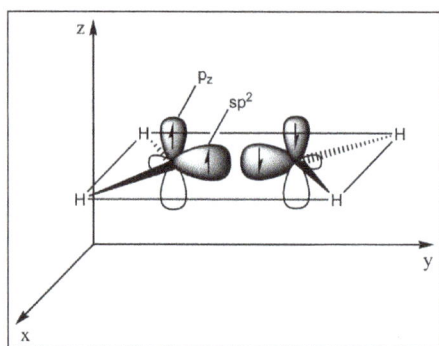

2 sp² orbitals (total of 3 such orbitals) approach to form a sp²-sp² sigma bond

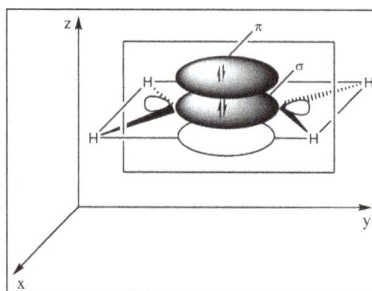

Two p-orbitals overlap to form a pi-bond in a plane parallel to the sigma plane

pi bond (green) in ethylene

With 133 pm, the ethylene C=C bond length is shorter than the C–C length in ethane with 154 pm. The double bond is also stronger, 636 kJ mol⁻¹ versus 368 kJ mol⁻¹ but not twice as much as the pi-bond is weaker than the sigma bond due to less effective pi-overlap.

In an alternative representation, the double bond results from two overlapping sp³ orbitals as in a bent bond.

Types of Double Bonds between Atoms

	C	O	N	S
C	alkene	carbonyl group	imine	thioketone, thial
O		dioxygen	nitroso compound	sulfoxide, sulfone, sulfinic acid, sulfonic acid
N			azo compound	
S				disulfur

Variations

In molecules, with alternating double bonds and single bonds, p-orbital overlap can exist over multiple atoms in a chain, giving rise to a conjugated system. Conjugation can be found in systems such as dienes and enones. In cyclic molecules, conjugation can lead to aromaticity. In cumulenes, two double bonds are adjacent.

Double bonds are common for period 2 elements carbon, nitrogen, and oxygen, and less common with elements of higher periods. Metals, too, can engage in multiple bonding in a metal ligand multiple bond.

Group 14 Alkene Homologs

Double bonded compounds, alkene homologs, $R_2E=ER_2$ are now known for all of the heavier group 14 elements. Unlike the alkenes these compounds are not planar but adopt twisted and/or trans bent structures. These effects become more pronounced for the heavier elements. The distannene $(Me_3Si)_2CHSn=SnCH(SiMe_3)_2$ has a tin-tin bond length just a little shorter than a single bond, a trans bent structure with pyramidal coordination at each tin atom, and readily dissociates in solution to form (Me$_3$Si)$_2$CHSn: (stannanediyl, a carbene analog). The bonding comprises two weak donor acceptor bonds, the lone pair on each tin atom overlapping with the empty p orbital on the other. In contrast, in disilenes each silicon atom has planar coordination but the substituents are twisted so that the molecule as a whole is not planar. In diplumbenes the Pb=Pb bond length can be longer than that of many corresponding single bonds Plumbenes and stannenes generally dissociate in solution into monomers with bond enthalpies that are just a fraction of the corresponding single bonds. Some double bonds plumbenes and stannenes are similar in strength to hydrogen bonds. The Carter-Goddard-Malrieu-Trinquier model can be used to predict the nature of the bonding.

Triple Bond

A triple bond in chemistry is a chemical bond between two atoms involving six bonding electrons instead of the usual two in a covalent single bond. The most common triple bond, that between two carbon atoms, can be found in alkynes. Other functional groups containing a triple bond are cyanides and isocyanides. Some diatomic molecules, such as dinitrogen and carbon monoxide, are also triple bonded. In skeletal formula the triple bond is drawn as three parallel lines (≡) between the two connected atoms.

Triple bonds are stronger than the equivalent single bonds or double bonds, with a bond order of three.

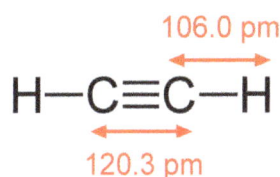

acetylene, H–C≡C–H

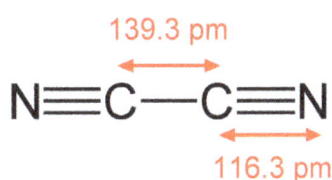

cyanogen, N≡C–C≡N

carbon monoxide, C≡O

Chemical compounds with triple bond

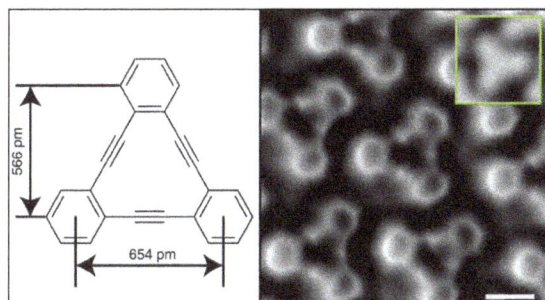

Structure and AFM image of dehydrobenzo annulene, where benzene rings are held together by triple bonds

Bonding

The types of bonding can be explained in terms of orbital hybridization. In the case of acetylene each carbon atom has two sp orbitals and two p-orbitals. The two sp orbitals are linear with 180°

angles and occupy the x-axis (cartesian coordinate system). The p-orbitals are perpendicular on the y-axis and the z-axis. When the carbon atoms approach each other, the sp orbitals overlap to form a sp-sp sigma bond. At the same time the p_z-orbitals approach and together they form a p_z-p_z pi-bond. Likewise, the other pair of p_y-orbitals form a p_y-p_y pi-bond. The result is formation of one sigma bond and two pi bonds.

In the bent bond model, the triple bond can also formed by the overlapping of three sp^3 lobes without the need to invoke a pi-bond.

References

- Kubas, Gregory (2002). "Metal Dihydrogen and σ-Bond Complexes: Structure, Theory, and Reactivity". J. Am. Chem. Soc. 124 (14): 3799–3800. Doi:10.1021/ja0153417

- Pi-bonds, chemistry: byjus.com, Retrieved 15 February, 2019

- Keeler, James; Wothers, Peter (May 2008). Chemical Structure and Reactivity (1st ed.). Oxford: OUP Oxford. pp. 27–46. ISBN 978-0199289301

- Wang, Yuzhong; Robinson, Gregory H. (2009). "Unique homonuclear multiple bonding in main group compounds". Chemical Communications. Royal Society of Chemistry (35): 5201–5213. doi:10.1039/B908048A

- Pyykkö, Pekka; Riedel, Sebastian; Patzschke, Michael (2005). "Triple-Bond Covalent Radii". Chemistry: A European Journal. 11 (12): 3511–20. doi:10.1002/chem.200401299. PMID 15832398

5

Theories in Quantum Chemistry

Some of the theories used in quantum chemistry are valence bond theory, molecular orbit theory, VSEPR theory, density functional theory, time-dependent density functional theory, etc. This chapter closely examines these theories of quantum chemistry to provide an extensive understanding of the subject.

Valence Bond Theory

Valence bond theory describes a covalent bond as the overlap of half-filled atomic orbitals (each containing a single electron) that yield a pair of electrons shared between the two bonded atoms. We say that orbitals on two different atoms overlap when a portion of one orbital and a portion of a second orbital occupy the same region of space. According to valence bond theory, a covalent bond results when two conditions are met: (1) an orbital on one atom overlaps an orbital on a second atom and (2) the single electrons in each orbital combine to form an electron pair. The mutual attraction between this negatively charged electron pair and the two atoms' positively charged nuclei serves to physically link the two atoms through a force we define as a covalent bond. The strength of a covalent bond depends on the extent of overlap of the orbitals involved. Orbitals that overlap extensively form bonds that are stronger than those that have less overlap.

The energy of the system depends on how much the orbitals overlap. Figure illustrates how the sum of the energies of two hydrogen atoms (the colored curve) changes as they approach each other. When the atoms are far apart there is no overlap, and by convention we set the sum of the energies at zero. As the atoms move together, their orbitals begin to overlap. Each electron begins to feel the attraction of the nucleus in the other atom. In addition, the electrons begin to repel each other, as do the nuclei. While the atoms are still widely separated, the attractions are slightly stronger than the repulsions, and the energy of the system decreases. (A bond begins to form). As the atoms move closer together, the overlap increases, so the attraction of the nuclei for the electrons continues to increase (as do the repulsions among electrons and between the nuclei). At some specific distance between the atoms, which varies depending on the atoms involved, the energy reaches its lowest (most stable) value. This optimum distance between the two bonded nuclei is the bond distance between the two atoms. The bond is stable because at this point, the attractive and repulsive forces combine to create the lowest possible energy configuration. If the distance between the nuclei were to decrease further, the repulsions between nuclei and the repulsions as electrons are confined in closer proximity to each other would become stronger than the attractive forces. The energy of the system would then rise (making the system destabilized), as shown at the far left of figure.

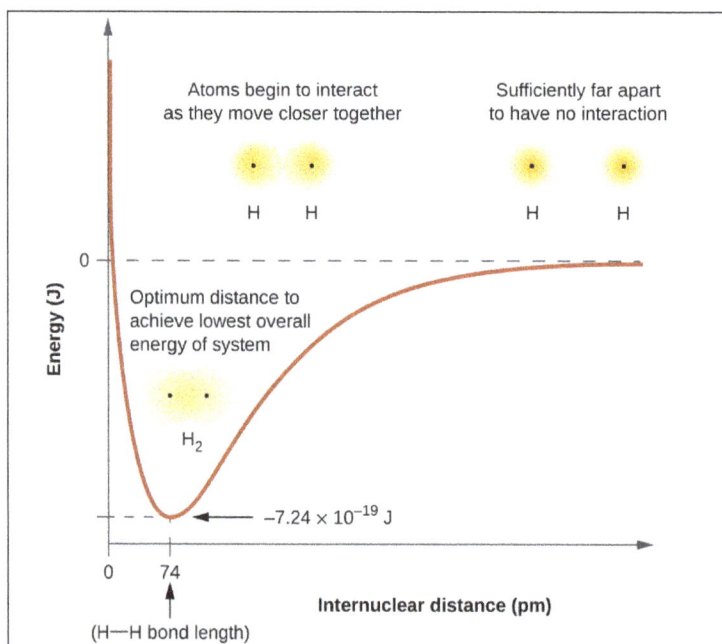

The interaction of two hydrogen atoms changes as a function of distance. (b) The energy of the system changes as the atoms interact. The lowest (most stable) energy occurs at a distance of 74 pm, which is the bond length observed for the H_2 molecule.

The bond energy is the difference between the energy minimum (which occurs at the bond distance) and the energy of the two separated atoms. This is the quantity of energy released when the bond is formed. Conversely, the same amount of energy is required to break the bond. For the H_2 molecule, at the bond distance of 74 pm the system is 7.24×10^{-19} J lower in energy than the two separated hydrogen atoms. This may seem like a small number. However, that bond energies are often discussed on a per-mole basis. For example, it requires 7.24×10^{-19} J to break one H–H bond, but it takes 4.36×10^5 J to break 1 mole of H–H bonds. A comparison of some bond lengths and energies is shown in Table. We can find many of these bonds in a variety of molecules, and this table provides average values. For example, breaking the first C–H bond in CH_4 requires 439.3 kJ/mol, while breaking the first C–H bond in $H-CH_2C_6H_5$ (a common paint thinner) requires 375.5 kJ/mol.

Table: Representative Bond Energies and Lengths

Bond	Length (pm)	Energy (kJ/mol)	Bond	Length (pm)	Energy (kJ/mol)
H–H	74	436	C–O	140.1	358
H–C	106.8	413	C=O	119.7	745
H–N	101.5	391	C≡O	113.7	1072
H–O	97.5	467	H–Cl	127.5	431
C–C	150.6	347	H–Br	141.4	366

Bond	Length (pm)	Energy (kJ/mol)	Bond	Length (pm)	Energy (kJ/mol)
C=C	133.5	614	H–I	160.9	298
C≡C	120.8	839	O–O	148	146
C–N	142.1	305	O=O	120.8	498
C=N	130.0	615	F–F	141.2	159
C≡N	116.1	891	Cl–Cl	198.8	243

In addition to the distance between two orbitals, the orientation of orbitals also affects their overlap (other than for two s orbitals, which are spherically symmetric). Greater overlap is possible when orbitals are oriented such that they overlap on a direct line between the two nuclei. Figure illustrates this for two p orbitals from different atoms; the overlap is greater when the orbitals overlap end to end rather than at an angle.

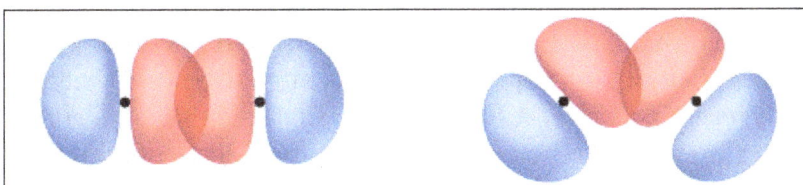

(a) The overlap of two p orbitals is greatest when the orbitals are directed end to end. (b) Any other arrangement results in less overlap. The dots indicate the locations of the nuclei.

The overlap of two s orbitals (as in H_2), the overlap of an s orbital and a p orbital (as in HCl), and the end-to-end overlap of two p orbitals (as in Cl_2) all produce sigma bonds (σ bonds), as illustrated in figure A σ bond is a covalent bond in which the electron density is concentrated in the region along the internuclear axis; that is, a line between the nuclei would pass through the center of the overlap region. Single bonds in Lewis structures are described as σ bonds in valence bond theory.

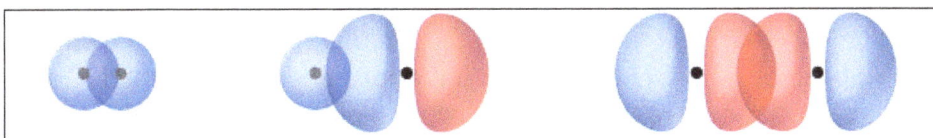

Sigma (σ) bonds form from the overlap of the following: (a) two s orbitals, (b) an s orbital and a p orbital, and (c) two p orbitals. The dots indicate the locations of the nuclei.

A pi bond (π bond) is a type of covalent bond that results from the side-by-side overlap of two porbitals, as illustrated in figure. In a π bond, the regions of orbital overlap lie on opposite sides of the internuclear axis. Along the axis itself, there is a node, that is, a plane with no probability of finding an electron.

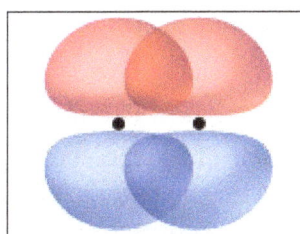

Pi (π) bonds form from the side-by-side overlap of two p orbitals. The dots indicate the location of the nuclei.

While all single bonds are σ bonds, multiple bonds consist of both σ and π bonds. As the Lewis structures in suggest, O_2 contains a double bond, and N_2 contains a triple bond. The double bond consists of one σ bond and one π bond, and the triple bond consists of one σ bond and two π bonds. Between any two atoms, the first bond formed will always be a σ bond, but there can only be one σ bond in any one location. In any multiple bond, there will be one σ bond, and the remaining one or two bonds will be π bonds.

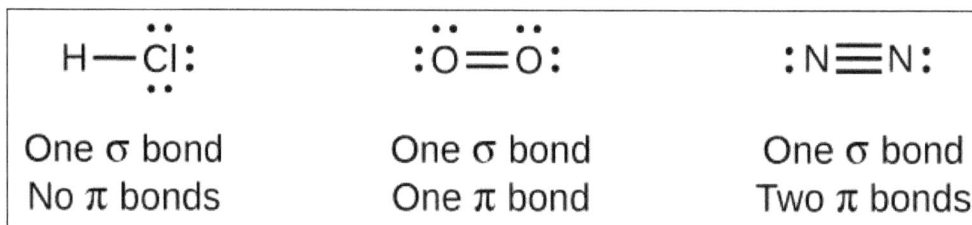

H—C̈l:	:Ö═Ö:	:N≡N:
One σ bond	One σ bond	One σ bond
No π bonds	One π bond	Two π bonds

As seen in Table, an average carbon-carbon single bond is 347 kJ/mol, while in a carbon-carbon double bond, the π bond increases the bond strength by 267 kJ/mol. Adding an additional π bond causes a further increase of 225 kJ/mol. We can see a similar pattern when we compare other σ and π bonds. Thus, each individual π bond is generally weaker than a corresponding σ bond between the same two atoms. In a σ bond, there is a greater degree of orbital overlap than in a π bond.

Molecular Orbital Theory

Because arguments based on atomic orbitals focus on the bonds formed between valence electrons on an atom, they are often said to involve a valence-bond theory.

The valence-bond model can't adequately explain the fact that some molecules contains two equivalent bonds with a bond order between that of a single bond and a double bond. The best it can do is suggest that these molecules are mixtures, or hybrids, of the two Lewis structures that can be written for these molecules.

This problem, and many others, can be overcome by using a more sophisticated model of bonding based on molecular orbitals. Molecular orbital theory is more powerful than valence-bond theory because the orbitals reflect the geometry of the molecule to which they are applied. But this power carries a significant cost in terms of the ease with which the model can be visualized.

Forming Molecular Orbitals

Molecular orbitals are obtained by combining the atomic orbitals on the atoms in the molecule. Consider the H_2 molecule, for example. One of the molecular orbitals in this molecule is constructed by adding the mathematical functions for the two 1s atomic orbitals that come together to form this molecule. Another orbital is formed by subtracting one of these functions from the other, as shown in the figure below.

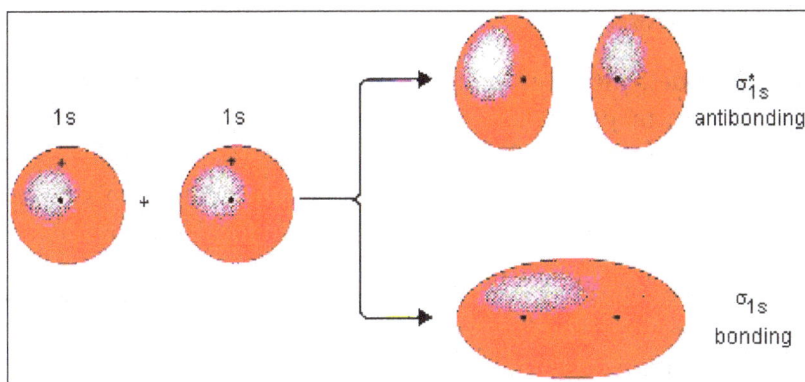

One of these orbitals is called a bonding molecular orbital because electrons in this orbital spend most of their time in the region directly between the two nuclei. It is called a sigma (σ) molecular orbital because it looks like an s orbital when viewed along the H-H bond. Electrons placed in the other orbital spend most of their time away from the region between the two nuclei. This orbital is therefore an antibonding, or sigma star (σ^*), molecular orbital.

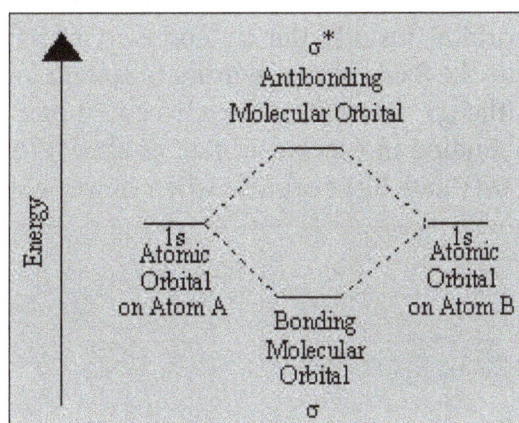

The σ bonding molecular orbital concentrates electrons in the region directly between the two nuclei. Placing an electron in this orbital therefore stabilizes the H_2 molecule. Since the σ^* antibonding molecular orbital forces the electron to spend most of its time away from the area between the nuclei, placing an electron in this orbital makes the molecule less stable.

Electrons are added to molecular orbitals, one at a time, starting with the lowest energy molecular orbital. The two electrons associated with a pair of hydrogen atoms are placed in the lowest energy, or σ bonding, molecular orbital, as shown in the figure below. This diagram suggests that the energy of an H_2 molecule is lower than that of a pair of isolated atoms. As a result, the H_2 molecule is more stable than a pair of isolated atoms.

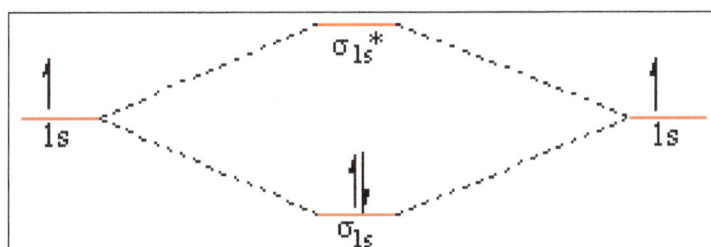

Why some Molecules do not Exist?

This molecular orbital model can be used to explain why He$_2$ molecules don't exist. Combining a pair of helium atoms with 1s^2 electron configurations would produce a molecule with a pair of electrons in both the σ bonding and the   * antibonding molecular orbitals. The total energy of an He$_2$ molecule would be essentially the same as the energy of a pair of isolated helium atoms, and there would be nothing to hold the helium atoms together to form a molecule.

The fact that an He$_2$ molecule is neither more nor less stable than a pair of isolated helium atoms illustrates an important principle: The core orbitals on an atom make no contribution to the stability of the molecules that contain this atom. The only orbitals that are important in our discussion of molecular orbitals are those formed when valence-shell orbitals are combined. The molecular orbital diagram for an O$_2$ molecule would therefore ignore the 1s electrons on both oxygen atoms and concentrate on the interactions between the 2s and 2p valence orbitals.

Molecular Orbitals of the Second Energy Level

The 2s orbitals on one atom combine with the 2s orbitals on another to form a σ$_{2s}$ bonding and a σ$_{2s}$* antibonding molecular orbital, just like the σ$_{1s}$ and σ$_{1s}$* orbitals formed from the 1s atomic orbitals. If we arbitrarily define the Zaxis of the coordinate system for the O$_2$ molecule as the axis along which the bond forms, the 2p$_z$ orbitals on the adjacent atoms will meet head-on to form a σ$_{2p}$ bonding and a σ$_{2p}$* antibonding molecular orbital, as shown in the figure below. These are called sigma orbitals because they look like s orbitals when viewed along the oxygen-oxygen bond.

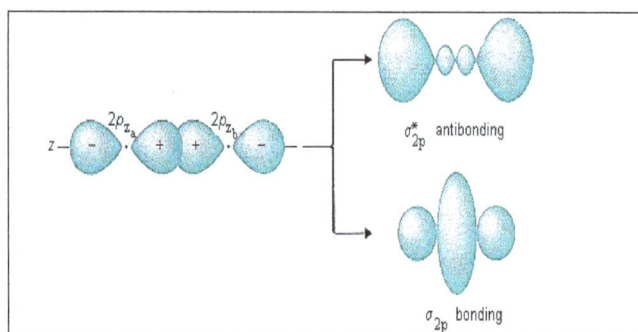

The 2p$_x$ orbitals on one atom interact with the 2p$_x$ orbitals on the other to form molecular orbitals that have a different shape, as shown in the figure below. These molecular orbitals are called *pi* (π) orbitals because they look like p orbitals when viewed along the bond. Whereas σ and σ* orbitals concentrate the electrons along the axis on which the nuclei of the atoms lie, π and π* orbitals concentrate the electrons either above or below this axis.

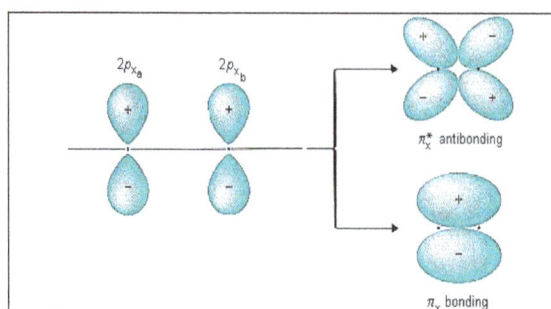

The $2p_x$ atomic orbitals combine to form a π_x bonding molecular orbital and a π_x^* antibonding molecular orbital. The same thing happens when the $2p_y$ orbitals interact, only in this case we get a π_y and a π_y^* antibonding molecular orbital. Because there is no difference between the energies of the $2p_x$ and $2p_y$ atomic orbitals, there is no difference between the energies of the π_x and π_y or the π_x^* and π_y^* molecular orbitals.

The interaction of four valence atomic orbitals on one atom ($2s$, $2p_x$, $2p_y$ and $2p_z$) with a set of four atomic orbitals on another atom leads to the formation of a total of eight molecular orbitals: $\sigma_{2s}, \sigma_{2s}^*, \sigma_{2p}, \sigma_{2p}^*, \pi_x, \pi_y, \pi_x^*,$ and π_y^*.

There is a significant difference between the energies of the $2s$ and $2p$ orbitals on an atom. As a result, the σ_{2s} and σ_{2s}^* orbitals both lie at lower energies than the $\sigma_{2p}, \sigma_{2p}^*, \pi_x, \pi_y, \pi_x^*,$ and π_y^* orbitals. To sort out the relative energies of the six molecular orbitals formed when the $2p$ atomic orbitals on a pair of atoms are combined, we need to understand the relationship between the strength of the interaction between a pair of orbitals and the relative energies of the molecular orbitals they form.

Because they meet head-on, the interaction between the $2p_z$ orbitals is stronger than the interaction between the $2p_x$ or $2p_y$ orbitals, which meet edge-on. As a result, the σ_{2p} orbital lies at a lower energy than the π_x and π_y orbitals, and the σ_{2p}^* orbital lies at higher energy than the and π_y^* orbitals, as shown in the figure below.

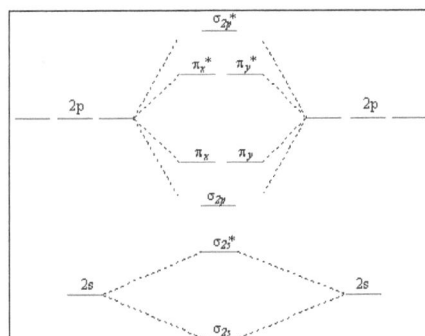

Unfortunately an interaction is missing from this model. It is possible for the $2s$ orbital on one atom to interact with the $2p_z$ orbital on the other. This interaction introduces an element of s-p mixing, or hybridization, into the molecular orbital theory. The result is a slight change in the relative energies of the molecular orbitals, to give the diagram shown in the figure below. Experiments have shown that O_2 and F_2 are best described by the model in the figure above, but B_2, C_2, and N_2 are best described by a model that includes hybridization, as shown in the figure below.

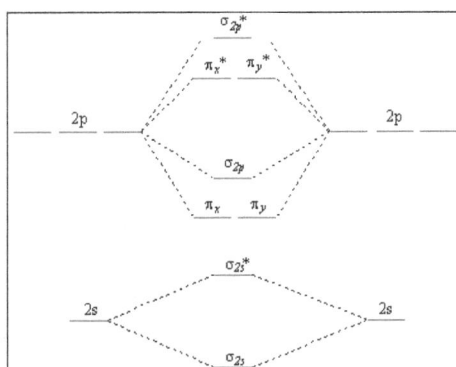

Bond Order

The number of bonds between a pair of atoms is called the bond order. Bond orders can be calculated from Lewis structures, which are the heart of the valence-bond model. Oxygen, for example, has a bond order of two.

$$:\overset{..}{O}\!=\!\overset{..}{O}:$$

When there is more than one Lewis structure for a molecule, the bond order is an average of these structures. The bond order in sulfur dioxide, for example, is 1.5 — the average of an S-O single bond in one Lewis structure and an S=O double bond in the other.

$$:\overset{..}{O}\!-\!\overset{..}{S}\!=\!\overset{..}{O}: \quad\longleftrightarrow\quad :\overset{..}{O}\!=\!\overset{..}{S}\!-\!\overset{..}{O}:$$

In molecular orbital theory, we calculate bond orders by assuming that two electrons in a bonding molecular orbital contribute one net bond and that two electrons in an antibonding molecular orbital cancel the effect of one bond. We can calculate the bond order in the O_2 molecule by noting that there are eight valence electrons in bonding molecular orbitals and four valence electrons in antibonding molecular orbitals in the electron configuration of this molecule. Thus, the bond order is two.

$$Bond\ order = \frac{bond\ in\ g\ eletrons - antibonding\ electrons}{2} = \frac{8-4}{2} = 2$$

Although the Lewis structure and molecular orbital models of oxygen yield the same bond order, there is an important difference between these models. The electrons in the Lewis structure are all paired, but there are two unpaired electrons in the molecular orbital description of the molecule. As a result, we can test the predictions of these theories by studying the effect of a magnetic field on oxygen.

Atoms or molecules in which the electrons are paired are diamagnetic — repelled by both poles of a magnetic. Those that have one or more unpaired electrons are paramagnetic — attracted to a magnetic field. Liquid oxygen is attracted to a magnetic field and can actually bridge the gap between the poles of a horseshoe magnet. The molecular orbital model of O_2 is therefore superior to the valence-bond model, which cannot explain this property of oxygen.

VSEPR Theory

Valence shell electron pair repulsion theory, or VSEPR theory, is a model used in chemistry to predict the geometry of individual molecules from the number of electron pairs surrounding their central atoms. It is also named the Gillespie-Nyholm theory after its two main developers, Ronald Gillespie and Ronald Nyholm. The premise of VSEPR is that the valence electron pairs surrounding an atom tend to repel each other and will, therefore, adopt an arrangement that minimizes this repulsion, thus determining the molecule's geometry. Gillespie has emphasized that the electron-electron repulsion due to the Pauli exclusion principle is more important in determining molecular geometry than the electrostatic repulsion.

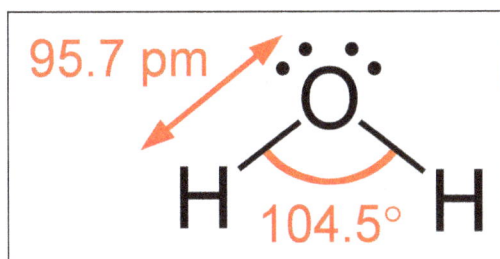

Example of bent electron arrangement. Shows location of unpaired electrons, bonded atoms, and bond angles. (Water molecule) The bond angle for water is 104.5°.

VSEPR theory is based on observable electron density rather than mathematical wave functions and hence unrelated to orbital hybridisation, although both address molecular shape. While it is mainly qualitative, VSEPR has a quantitative basis in quantum chemical topology (QCT) methods such as the electron localization function (ELF) and the quantum theory of atoms in molecules (QTAIM).

The idea of a correlation between molecular geometry and number of valence electron pairs (both shared and unshared pairs) was originally proposed in 1939 by Ryutaro Tsuchida in Japan, and was independently presented in a Bakerian Lecture in 1940 by Nevil Sidgwick and Herbert Powell of the University of Oxford. In 1957, Ronald Gillespie and Ronald Sydney Nyholm of University College London refined this concept into a more detailed theory, capable of choosing between various alternative geometries.

In recent years, VSEPR theory has been criticized as an outdated model from the standpoint of both scientific accuracy and pedagogical value. In particular, the equivalent lone pairs of water and carbonyl compounds in VSEPR theory neglect fundamental differences in the symmetries (σ vs. π) of molecular orbitals and natural bond orbitals that correspond to them, a difference that is sometimes chemically significant. Furthermore, there is little evidence, computational or experimental, proposing that lone pairs are "bigger" than bonding pairs. It has been suggested that Bent's rule is capable of replacing VSEPR as a simple model for explaining molecular structure. Nevertheless, VSEPR theory captures many of the essential features of the structure and electron distribution of simple molecules, and most undergraduate general chemistry courses continue to teach it.

VSEPR theory is used to predict the arrangement of electron pairs around non-hydrogen atoms in molecules, especially simple and symmetric molecules, where these key, central atoms participate in bonding to two or more other atoms; the geometry of these key atoms and their non-bonding electron pairs in turn determine the geometry of the larger whole.

The number of electron pairs in the valence shell of a central atom is determined after drawing the Lewis structure of the molecule, and expanding it to show all bonding groups and lone pairs of electrons. In VSEPR theory, a double bond or triple bond are treated as a single bonding group. The sum of the number of atoms bonded to a central atom and the number of lone pairs formed by its nonbonding valence electrons is known as the central atom's steric number.

The electron pairs (or groups if multiple bonds are present) are assumed to lie on the surface of a sphere centered on the central atom and tend to occupy positions that minimize their mutual repulsions by maximizing the distance between them. The number of electron pairs (or groups), therefore, determines the overall geometry that they will adopt. For example, when there are two electron pairs surrounding the central atom, their mutual repulsion is minimal when they lie at opposite poles of the sphere. Therefore, the central atom is predicted to adopt a linear geometry. If there are 3 electron pairs surrounding the central atom, their repulsion is minimized by placing them at the vertices of an equilateral triangle centered on the atom. Therefore, the predicted geometry is trigonal. Likewise, for 4 electron pairs, the optimal arrangement is tetrahedral.

Degree of Repulsion

The overall geometry is further refined by distinguishing between bonding and nonbonding electron pairs. The bonding electron pair shared in a sigma bond with an adjacent atom lies further from the central atom than a nonbonding (lone) pair of that atom, which is held close to its positively charged nucleus. VSEPR theory therefore views repulsion by the lone pair to be greater than the repulsion by a bonding pair. As such, when a molecule has 2 interactions with different degrees of repulsion, VSEPR theory predicts the structure where lone pairs occupy positions that allow them to experience less repulsion. Lone pair–lone pair (lp–lp) repulsions are considered stronger than lone pair–bonding pair (lp–bp) repulsions, which in turn are considered stronger than bonding pair–bonding pair (bp–bp) repulsions, distinctions that then guide decisions about overall geometry when 2 or more non-equivalent positions are possible. For instance, when 5 valence electron pairs surround a central atom, they adopt a trigonal bipyramidal molecular geometry with two collinear axial positions and three equatorial positions. An electron pair in an axial position has three close equatorial neighbors only 90° away and a fourth much farther at 180°, while an equatorial electron pair has only two adjacent pairs at 90° and two at 120°. The repulsion from the close neighbors at 90° is more important, so that the axial positions experience more repulsion than the equatorial positions; hence, when there are lone pairs, they tend to occupy equatorial positions as shown in the diagrams of the next section for steric number five.

The difference between lone pairs and bonding pairs may also be used to rationalize deviations from idealized geometries. For example, the H_2O molecule has four electron pairs in its valence shell: two lone pairs and two bond pairs. The four electron pairs are spread so as to point roughly towards the apices of a tetrahedron. However, the bond angle between the two O–H bonds is only 104.5°, rather than the 109.5° of a regular tetrahedron, because the two lone pairs (whose density or probability envelopes lie closer to the oxygen nucleus) exert a greater mutual repulsion than the two bond pairs.

A bond of higher bond order also exerts greater repulsion since the pi bond electrons contribute.. For example in isobutylene, $(H_3C)_2C=CH_2$, the H_3C–C=C angle (124°) is larger than the H_3C–C–CH_3 angle (111.5°). However, in the carbonate ion, CO_3^{2-}, all three C–O bonds are equivalent with angles of 120° due to resonance.

AXE Method

The "AXE method" of electron counting is commonly used when applying the VSEPR theory. The electron pairs around a central atom are represented by a formula AX_nE_m, where A represents the central atom and always has an implied subscript one. Each X represents a ligand (an atom bonded to A). Each E represents a lone pair of electrons on the central atom. The total number of X and E is known as the steric number. For example in a molecule AX_3E_2, the atom A has a steric number of 5.

Based on the steric number and distribution of Xs and Es, VSEPR theory makes the predictions in the following tables. Note that the geometries are named according to the atomic positions only and not the electron arrangement. For example, the description of AX_2E_1 as a bent molecule means that the three atoms AX_2 are not in one straight line, although the lone pair helps to determine the geometry.

When the substituent (X) atoms are not all the same, the geometry is still approximately valid, but the bond angles may be slightly different from the ones where all the outside atoms are the same. For example, the double-bond carbons in alkenes like C_2H_4 are AX_3E_0, but the bond angles are not all exactly 120°. Likewise, $SOCl_2$ is AX_3E_1, but because the X substituents are not identical, the X–A–X angles are not all equal.

As a tool in predicting the geometry adopted with a given number of electron pairs, an often used physical demonstration of the principle of minimal electron pair repulsion utilizes inflated balloons. Through handling, balloons acquire a slight surface electrostatic charge that results in the adoption of roughly the same geometries when they are tied together at their stems as the corresponding number of electron pairs. For example, five balloons tied together adopt the trigonal bipyramidal geometry, just as do the five bonding pairs of a PCl_5 molecule (AX_5) or the two bonding and three non-bonding pairs of a XeF_2 molecule (AX_2E_3). The molecular geometry of the former is also trigonal bipyramidal, whereas that of the latter is linear.

Possible geometries for steric numbers of 10, 11, 12, or 14 are bicapped square antiprismatic (or bicapped dodecadeltahedral), octadecahedral, icosahedral, and bicapped hexagonal antiprismatic, respectively. No compounds with steric numbers this high involving monodentate ligands exist, and those involving multidentate ligands can often be analysed more simply as complexes with lower steric numbers when some multidentate ligands are treated as a unit.

Examples:

The methane molecule (CH_4) is tetrahedral because there are four pairs of electrons. The four hydrogen atoms are positioned at the vertices of a tetrahedron, and the bond angle is $\cos^{-1}(-\frac{1}{3})$ ≈ 109° 28′. This is referred to as an AX_4 type of molecule. As mentioned above, A represents the central atom and X represents an outer atom.

The ammonia molecule (NH_3) has three pairs of electrons involved in bonding, but there is a lone pair of electrons on the nitrogen atom. It is not bonded with another atom; however, it influences the overall shape through repulsions. As in methane above, there are four regions of electron density. Therefore, the overall orientation of the regions of electron density is tetrahedral. On the other hand, there are only three outer atoms. This is referred to as an AX_3E type molecule because the lone pair is represented by an E. By definition, the molecular shape or geometry describes the geometric arrangement of the atomic nuclei only, which is trigonal-pyramidal for NH_3.

Steric numbers of 7 or greater are possible, but are less common. The steric number of 7 occurs in iodine heptafluoride (IF_7); the base geometry for a steric number of 7 is pentagonal bipyramidal. The most common geometry for a steric number of 8 is a square antiprismatic geometry. Examples of this include the octacyanomolybdate ($Mo(CN)_8^{4-}$) and octafluorozirconate (ZrF_8^{4-}) anions. The nonahydridorhenate ion (ReH_9^{2-}) in potassium nonahydridorhenate is a rare example of a compound with a steric number of 9, which has a tricapped trigonal prismatic geometry.

Exceptions

There are groups of compounds where VSEPR fails to predict the correct geometry.

Some AX_2E_0 Molecules

The gas phase structures of the triatomic halides of the heavier members of group 2, (i.e., calcium, strontium and barium halides, MX_2), are not linear as predicted but are bent, (approximate X−M−X angles: CaF_2, 145°; SrF_2, 120°; BaF_2, 108°; $SrCl_2$, 130°; $BaCl_2$, 115°; $BaBr_2$, 115°; BaI_2, 105°). It has been proposed by Gillespie that this is caused by interaction of the ligands with the electron core of the metal atom, polarising it so that the inner shell is not spherically symmetric, thus influencing the molecular geometry. Ab initio calculations have been cited to propose that contributions from d orbitals in the shell below the valence shell are responsible, together with the overlap of other orbitals. Disilynes are also bent, despite having no lone pairs.

Some AX_2E_2 Molecules

One example of the AX_2E_2 geometry is molecular lithium oxide, Li_2O, a linear rather than bent structure, which is ascribed to its bonds being essentially ionic and the strong lithium-lithium repulsion that results. Another example is $O(SiH_3)_2$ with an Si−O−Si angle of 144.1°, which compares to the angles in Cl_2O (110.9°), $(CH_3)_2O$ (111.7°), and $N(CH_3)_3$ (110.9°). Gillespie and Robinson rationalize the Si−O−Si bond angle based on the observed ability of a ligand's lone pair to most greatly repel other electron pairs when the ligand electronegativity is greater than or equal to that of the central atom. In $O(SiH_3)_2$, the central atom is more electronegative, and the lone pairs are less localized and more weakly repulsive. The larger Si−O−Si bond angle results from this and strong ligand-ligand repulsion by the relatively large -SiH_3 ligand. Burford et al showed through X-ray diffraction studies that $Cl_3Al−O−PCl_3$ has a linear Al−O−P bond angle and is therefore a non-VSEPR molecule.

Some AX_6E_1 and AX_8E_1 Molecules

Xenon hexafluoride, which has a distorted octahedral geometry.

Some AX_6E_1 molecules, e.g. xenon hexafluoride (XeF_6) and the Te(IV) and Bi(III) anions, $TeCl_6^{2-}$, $TeBr_6^{2-}$, $BiCl_6^{3-}$, $BiBr_6^{3-}$ and BiI_6^{3-}, are octahedra, rather than pentagonal pyramids, and the lone pair does not affect the geometry to the degree predicted by VSEPR. Similarly, the octafluoroxenate ion (XeF_8^{2-}) in nitrosonium octafluoroxenate(VI) is a square antiprism and not a bicapped trigonal prism (as predicted by VSEPR theory for an AX_8E_1 molecule), despite having a lone pair. One rationalization is that steric crowding of the ligands allows little or no room for the non-bonding lone pair; another rationalization is the inert pair effect.

Transition Metal Molecules

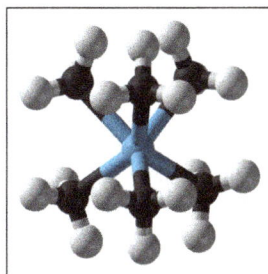

Hexamethyltungsten, a transition metal compound whose geometry is different from main group coordination.

Many transition metal compounds have unusual geometries, which can be ascribed to ligand bonding interaction with the d subshell and to absence of valence shell lone pairs. Gillespie suggested that this interaction can be weak or strong. Weak interaction is dealt with by the Kepert model, while strong interaction produces bonding pairs that also occupy the respective antipodal points of the sphere. This is similar to predictions based on sd hybrid orbitals using the VALBOND theory. The repulsion of these bidirectional bonding pairs leads to a different prediction of shapes.

Molecule type	Shape	Geometry	Examples
AX_2	Bent		VO_2^+
AX_3	Trigonal pyramidal		CrO_3
AX_4	Tetrahedral		$TiCl_4$
AX_5	Square pyramidal		$Ta(CH_3)_5$

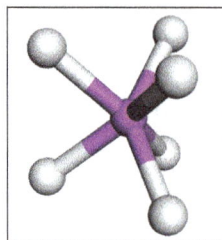

AX$_6$ C$_{3v}$ Trigonal prismatic   W(CH$_3$)$_6$

The Kepert model cannot explain the formation of square planar complexes.

Odd-Electron Molecules

The VSEPR theory can be extended to molecules with an odd number of electrons by treating the unpaired electron as a "half electron pair" - for example, Gillespie and Nyholm suggested that the decrease in the bond angle in the series $NO_2^+\left(180^\circ\right)$, $NO_2\left(134^\circ\right)$, $NO_2^-\left(115^\circ\right)$ indicates that a given set of bonding electron pairs exert a weaker repulsion on a single non-bonding electron than on a pair of non-bonding electrons. In effect, they considered nitrogen dioxide as an $AX_2E_{0.5}$ molecule, with a geometry intermediate between NO_2^+ and NO_2^-. Similarly, chlorine dioxide (ClO_2) is an $AX_2E_{1.5}$ molecule, with a geometry intermediate between ClO_2^- and ClO_2^-.

Finally, the methyl radical (CH_3) is predicted to be trigonal pyramidal like the methyl anion CH_3^-, but with a larger bond angle (as in the trigonal planar methyl cation (CH)). However, in this case, the VSEPR prediction is not quite true, as CH_3 is actually planar, although its distortion to a pyramidal geometry requires very little energy.

Density Functional Theory

Density functional theory (DFT) is a computational quantum mechanical modelling method used in physics, chemistry and materials science to investigate the electronic structure (or nuclear structure) (principally the ground state) of many-body systems, in particular atoms, molecules, and the condensed phases. Using this theory, the properties of a many-electron system can be determined by using functionals, i.e. functions of another function, which in this case is the spatially dependent electron density. Hence the name density functional theory comes from the use of functionals of the electron density. DFT is among the most popular and versatile methods available in condensed-matter physics, computational physics, and computational chemistry.

DFT has been very popular for calculations in solid-state physics since the 1970s. However, DFT was not considered accurate enough for calculations in quantum chemistry until the 1990s, when the approximations used in the theory were greatly refined to better model the exchange and correlation interactions. Computational costs are relatively low when compared to traditional methods, such as exchange only Hartree–Fock theory and its descendants that include electron correlation.

Despite recent improvements, there are still difficulties in using density functional theory to properly describe: intermolecular interactions (of critical importance to understanding chemical reactions), especially van der Waals forces (dispersion); charge transfer excitations; transition states, global potential energy surfaces, dopant interactions and some strongly correlated systems; and in calculations of the band gap and ferromagnetism in semiconductors. The incomplete treatment of dispersion can adversely affect the accuracy of DFT (at least when used alone and uncorrected) in the treatment of systems which are dominated by dispersion (e.g. interacting noble gas atoms) or where dispersion competes significantly with other effects (e.g. in biomolecules). The development of new DFT methods designed to overcome this problem, by alterations to the functional or by the inclusion of additive terms, is a current research topic.

Method

In the context of computational materials science, *ab initio* (from first principles) DFT calculations allow the prediction and calculation of material behaviour on the basis of quantum mechanical considerations, without requiring higher order parameters such as fundamental material properties. In contemporary DFT techniques the electronic structure is evaluated using a potential acting on the system's electrons. This DFT potential is constructed as the sum of external potentials V_{ext}, which is determined solely by the structure and the elemental composition of the system, and an effective potential V_{eff}, which represents interelectronic interactions. Thus, a problem for a representative supercell of a material with n electrons can be studied as a set of n one-electron Schrödinger-like equations, which are also known as Kohn–Sham equations.

Although density functional theory has its roots in the Thomas–Fermi model for the electronic structure of materials, DFT was first put on a firm theoretical footing by Walter Kohn and Pierre Hohenberg in the framework of the two Hohenberg–Kohn theorems (H–K). The original H–K theorems held only for non-degenerate ground states in the absence of a magnetic field, although they have since been generalized to encompass these.

The first H–K theorem demonstrates that the ground state properties of a many-electron system are uniquely determined by an electron density that depends on only three spatial coordinates. It set down the groundwork for reducing the many-body problem of N electrons with $3N$ spatial coordinates to three spatial coordinates, through the use of functionals of the electron density. This theorem has since been extended to the time-dependent domain to develop time-dependent density functional theory (TDDFT), which can be used to describe excited states.

The second H–K theorem defines an energy functional for the system and proves that the correct ground state electron density minimizes this energy functional.

In work that later won them the Nobel prize in chemistry, the H–K theorem was further developed by Walter Kohn and Lu Jeu Sham to produce Kohn–Sham DFT (KS DFT). Within this framework, the intractable many-body problem of interacting electrons in a static external potential is reduced to a tractable problem of noninteracting electrons moving in an effective potential. The effective potential includes the external potential and the effects of the Coulomb interactions between the electrons, e.g., the exchange and correlation interactions. Modeling the latter two interactions becomes the difficulty within KS DFT. The simplest approximation is the local-density approximation (LDA), which is based upon exact exchange energy for a uniform electron gas, which can be obtained from the Thomas–Fermi model, and from fits to the correlation energy for a uniform electron gas. Non-interacting systems are relatively easy to solve as the wavefunction can be represented as a Slater determinant of orbitals. Further, the kinetic energy functional of such a system is known exactly. The exchange–correlation part of the total energy functional remains unknown and must be approximated.

Another approach, less popular than KS DFT but arguably more closely related to the spirit of the original H–K theorems, is orbital-free density functional theory (OFDFT), in which approximate functionals are also used for the kinetic energy of the noninteracting system.

Derivation and Formalism

As usual in many-body electronic structure calculations, the nuclei of the treated molecules or clusters are seen as fixed (the Born–Oppenheimer approximation), generating a static external potential V in which the electrons are moving. A stationary electronic state is then described by a wavefunction $\Psi(r_1\rightarrow,...,r_N\rightarrow)$ satisfying the many-electron time-independent Schrödinger equation,

$$\hat{H}\varnothing = \left[\hat{T}+\hat{V}+\hat{U}\right]\varnothing = \left[\sum_i^N\left(-\frac{\hbar^2}{2m_i}\nabla_i^2\right)+\sum_i^N V\left(\vec{r}_i\right)+\sum_{i<j}^N U\left(\vec{r}_i,\vec{r}_j\right)\right]\varnothing = E\varnothing$$

where, for the N-electron system, $\hat{H}$ is the Hamiltonian, E is the total energy, $\hat{T}$ is the kinetic energy, $\hat{V}$ is the potential energy from the external field due to positively charged nuclei, and $\hat{U}$ is the electron–electron interaction energy. The operators $\hat{T}$ and $\hat{U}$ are called universal operators as they are the same for any N-electron system, while $\hat{V}$ is system-dependent. This complicated many-particle equation is not separable into simpler single-particle equations because of the interaction term $\hat{U}$.

There are many sophisticated methods for solving the many-body Schrödinger equation based on the expansion of the wavefunction in Slater determinants. While the simplest one is the Hartree–Fock method, more sophisticated approaches are usually categorized as post-Hartree–Fock methods. However, the problem with these methods is the huge computational effort, which makes it virtually impossible to apply them efficiently to larger, more complex systems.

Here DFT provides an appealing alternative, being much more versatile as it provides a way to systematically map the many-body problem, with $\hat{U}$, onto a single-body problem without $\hat{U}$. In DFT the key variable is the electron density $n(\vec{r})$, which for a normalized Ψ is given by:

$$n(\vec{r}) = N\int d^3r_2 \cdots \int d^3r_N\, \Psi^*\left(\vec{r},\vec{r}_2,...,\vec{r}_N\right)\Psi\left(\vec{r},\vec{r}_2,...,\vec{r}_N\right).$$

This relation can be reversed, i.e., for a given ground-state density $n_0(\vec{r})$ it is possible, in principle, to calculate the corresponding ground-state wavefunction $\Psi_0(\vec{r}_1,...,\vec{r}_N)$. In other words, Ψ is a unique functional of n_0:

$$\Psi_0 = \Psi[n_0]$$
,

and consequently the ground-state expectation value of an observable $\hat{O}$ is also a functional of n_0:

$$O[n_0] = \left\langle \Psi[n_0] \middle| \hat{O} \middle| \Psi[n_0] \right\rangle.$$

In particular, the ground-state energy is a functional of n_0:

$$E_0 = E[n_0] = \left\langle \Psi[n_0] \middle| \hat{T} + \hat{V} + \hat{U} \middle| \Psi[n_0] \right\rangle,$$

where the contribution of the external potential $\langle \Psi[n_0] | \hat{V} | \Psi[n_0] \rangle$ can be written explicitly in terms of the ground-state density n_0:

$$V[n_0] = \int V(\vec{r}) n_0(\vec{r}) \mathrm{d}^3 r.$$

More generally, the contribution of the external potential $\langle \Psi | \hat{V} | \Psi \rangle$ can be written explicitly in terms of the density n:

$$V[n] = \int V(\vec{r}) n(\vec{r}) \mathrm{d}^3 r.$$

The functionals $T[n]$ and $U[n]$ are called universal functionals, while $V[n]$ is called a non-universal functional, as it depends on the system under study. Having specified a system, i.e., having specified $\hat{V}$, one then has to minimize the functional

$$E[n] = T[n] + U[n] + \int V(\vec{r}) n(\vec{r}) \mathrm{d}^3 r$$

with respect to $\overrightarrow{n(r)}$, assuming one has reliable expressions for $T[n]$ and $U[n]$. A successful minimization of the energy functional will yield the ground-state density n_0 and thus all other ground-state observables.

The variational problems of minimizing the energy functional $E[n]$ can be solved by applying the Lagrangian method of undetermined multipliers. First, one considers an energy functional that does not explicitly have an electron–electron interaction energy term,

$$E_s[n] = \left\langle \Psi_s[n] \middle| \hat{T} + \hat{V}_s \middle| \Psi_s[n] \right\rangle,$$

where $\hat{T}$ denotes the kinetic energy operator and $\hat{V}_s$ is an external effective potential in which the particles are moving, so that $\overrightarrow{n_s(r)} \stackrel{\text{def}}{=} \overrightarrow{n(r)}$.

Thus, one can solve the so-called Kohn–Sham equations of this auxiliary noninteracting system,

$$\left[-\frac{\hbar^2}{2m} \nabla^2 + V_s(\vec{r}) \right] \varphi_i(\vec{r}) = \varepsilon_i \varphi_i(\vec{r})$$

which yields the orbitals φ_i that reproduce the density $\overrightarrow{n(r)}$ of the original many-body system:

$$n(\vec{r}) \stackrel{\text{def}}{=} n_s(\vec{r}) = \sum_i^N \left| \varphi_i(\vec{r}) \right|^2.$$

The effective single-particle potential can be written in more detail as:

$$V_s(\vec{r}) = V(\vec{r}) + \int \frac{e^2 n_s(\vec{r}')}{|\vec{r} - \vec{r}'|} d^3 r' + V_{XC}[n_s(\vec{r})]$$

where the second term denotes the so-called Hartree term describing the electron–electron Coulomb repulsion, while the last term V_{XC} is called the exchange–correlation potential. Here, V_{XC} includes all the many-particle interactions. Since the Hartree term and V_{XC} depend on $n(r)$, which depends on the φ_i, which in turn depend on V_s, the problem of solving the Kohn–Sham equation has to be done in a self-consistent (i.e., iterative) way. Usually one starts with an initial guess for $n(r)$, then calculates the corresponding V_s and solves the Kohn–Sham equations for the φ_i. From these one calculates a new density and starts again. This procedure is then repeated until convergence is reached. A non-iterative approximate formulation called Harris functional DFT is an alternative approach to this.

- The one-to-one correspondence between electron density and single-particle potential is not so smooth. It contains kinds of non-analytic structure. $E_s[n]$ contains kinds of singularities, cuts and branches. This may indicate a limitation of our hope for representing exchange–correlation functional in a simple analytic form.

- It is possible to extend the DFT idea to the case of the Green function G instead of the density n. It is called as Luttinger–Ward functional (or kinds of similar functionals), written as $E[G]$. However, G is determined not as its minimum, but as its extremum. Thus we may have some theoretical and practical difficulties.

- There is no one-to-one correspondence between one-body density matrix $n(\vec{r}, \vec{r}')$ and the one-body potential $V(\vec{r}, \vec{r}')$. (Remember that all the eigenvalues of $n(\vec{r}, \vec{r}')$ are 1). In other words, it ends up with a theory similar to the Hartree–Fock (or hybrid) theory.

Relativistic Density Functional Theory (Ab Initio Functional Forms)

The same theorems can be proven in the case of relativistic electrons, thereby providing generalization of DFT for the relativistic case. Unlike the nonrelativistic theory, in the relativistic case it is possible to derive a few exact and explicit formulas for the relativistic density functional.

Let one consider an electron in a hydrogen-like ion obeying the relativistic Dirac equation. The Hamiltonian H for a relativistic electron moving in the Coulomb potential can be chosen in the following form (atomic units are used):

$$H = c(\vec{\alpha} \cdot \vec{p}) + eV + mc^2 \beta,$$

where $V = -\dfrac{eZ}{r}$ is the Coulomb potential of a pointlike nucleus, $\vec{p}$ is a momentum operator of the electron, and e, m and c are the elementary charge, electron mass and the speed of light respectively, and finally $\vec{\alpha}$ and β are a set of Dirac 2×2 matrices:

$$\vec{\alpha} = \begin{pmatrix} 0 & \vec{\sigma} \\ \vec{\sigma} & 0 \end{pmatrix},$$

$$\beta = \begin{pmatrix} I & 0 \\ 0 & -I \end{pmatrix}.$$

To find out the eigenfunctions and corresponding energies, one solves the eigenfunction equation:

$$H\Psi = E\Psi,$$

where $\Psi = (\Psi(1), \Psi(2), \Psi(3), \Psi(4))^{\mathrm{T}}$ is a four-component wavefunction and E is the associated eigenenergy. It is demonstrated in Brack that application of the virial theorem to the eigenfunction equation produces the following formula for the eigenenergy of any bound state:

$$E = mc^2 \langle \emptyset | \beta | \emptyset \rangle = mc^2 \int |\emptyset(1)|^2 + |\emptyset(2)|^2 - |\emptyset(3)|^2 - |\emptyset(4)|^2 \, d\tau,$$

and analogously, the virial theorem applied to the eigenfunction equation with the square of the Hamiltonian yields:

$$E^2 = m^2 c^4 + emc^2 \langle \Psi | V\beta | \Psi \rangle.$$

It is easy to see that both of the above formulae represent density functionals. The former formula can be easily generalized for the multi-electron case.

One may observe that both of the functionals written above don't have extremals, of course if reasonably wide set of functions is allowed for variation. Nevertheless it is possible to design a density functional with desired extremal properties out of those ones. Let us make it in the following way:

$$F[n] = \frac{1}{mc^2}\left(mc^2 \int n d\tau - \sqrt{m^2 c^4 + emc^2 \int Vn d\tau}\right)^2 + \delta_{n,n_e} mc^2 \int n d\tau,$$

where n_e in Kronecker delta symbol of the second term denotes any extremal for the functional represented by the first term of the functional F. The second term amounts to zero for any function which is not an extremal for the first term of functional F. To proceed further we'd like to find Lagrange equation for this functional. In order to do this we should allocate a linear part of functional increment when argument function is altered.

$$F[n_e + \delta n] = \frac{1}{mc^2}\left(mc^2 \int (n_e + \delta n)d\tau - \sqrt{m^2 c^4 + emc^2 \int V(n_e + \delta n)d\tau}\right)^2$$

Deploying written above equation it is easy to find the following formula for functional derivative:

$$\frac{\delta F[n_e]}{\delta n} = 2A - \frac{2B^2 + AeV(\tau_0)}{B} + eV(\tau_0),$$

where A and B stay for $mc^2\int n_e \, d\tau$ and $\sqrt{m^2c^4 + emc^2\int Vn_e d\tau}$ respectively. And finally $V(\tau_o)$ is a value of potential in some point, specified by support of variation function δn which is supposed to be infinitesimal. To advance toward Lagrange equation we equate functional derivative to zero and after simple algebraic manipulations arrive to the following equation.

$$2B(A - B) = eV(\tau_0)(A - B)$$

Apparently this equation could have solution only if A is equal to B. This last condition provides us with Lagrange equation for functional F, which could be finally written down in the following form.

$$\left(mc^2\int nd\tau\right)^2 = m^2c^4 + emc^2\int Vnd\tau$$

Solutions of this equation represent extremals for functional F. It's easy to see that all real densities, that is densities corresponding to the bound states of the system in question, are solutions of written above equation, which could be called as well Kohn-Sham equation in this particular case. Looking back onto the definition of the functional F we clearly see that the functional produces energy of the system for appropriate density, because the first term amounts to zero for such density and the second one delivers the energy value.

Exchange–correlation Functionals

The major problem with DFT is that the exact functionals for exchange and correlation are not known except for the free electron gas. However, approximations exist which permit the calculation of certain physical quantities quite accurately. One of the simplest approximations is the local-density approximation (LDA), where the functional depends only on the density at the coordinate where the functional is evaluated:

$$E_{XC}^{LDA}[n] = \int \varepsilon_{XC}(n)n(\vec{r})d^3r.$$

The local spin-density approximation (LSDA) is a straightforward generalization of the LDA to include electron spin:

$$E_{XC}^{LSDA}\left[n_\uparrow, n_\downarrow\right] = \int \varepsilon_{XC}\left(n_\uparrow, n_\downarrow\right)n(\vec{r})d^3r.$$

In LDA, the exchange–correlation energy is typically separated into the exchange part and the correlation part: $\varepsilon_{XC} = \varepsilon_X + \varepsilon_C$. The exchange part is called the Dirac (or sometimes Slater) exchange which takes the form $\varepsilon_X \propto n^{1/3}$. There are, however, many mathematical forms for the correlation part. Highly accurate formulae for the correlation energy density $\varepsilon_C\left(n_\uparrow, n_\downarrow\right)$ have been constructed from quantum Monte Carlo simulations of jellium. A simple first-principles correlation functional has been recently proposed as well. Although unrelated to the Monte Carlo simulation, the two variants provide comparable accuracy.

The LDA assumes that the density is the same everywhere. Because of this, the LDA has a tendency to underestimate the exchange energy and over-estimate the correlation energy. The errors due to the exchange and correlation parts tend to compensate each other to a certain degree. To correct for this tendency, it is common to expand in terms of the gradient of the density in order to account for the non-homogeneity of the true electron density. This allows for corrections based on the changes in density away from the coordinate. These expansions are referred to as generalized gradient approximations (GGA) and have the following form:

$$E_{XC}^{GGA}\left[n_\uparrow, n_\downarrow\right] = \int \varepsilon_{XC}\left(n_\uparrow, n_\downarrow, \nabla n_\uparrow, \nabla n_\downarrow\right)n(\vec{r})d^3r.$$

Using the latter (GGA), very good results for molecular geometries and ground-state energies have been achieved.

Potentially more accurate than the GGA functionals are the meta-GGA functionals, a natural development after the GGA (generalized gradient approximation). Meta-GGA DFT functional in its original form includes the second derivative of the electron density (the Laplacian) whereas GGA includes only the density and its first derivative in the exchange–correlation potential.

Functionals of this type are, for example, TPSS and the Minnesota Functionals. These functionals include a further term in the expansion, depending on the density, the gradient of the density and the Laplacian (second derivative) of the density.

Difficulties in expressing the exchange part of the energy can be relieved by including a component of the exact exchange energy calculated from Hartree–Fock theory. Functionals of this type are known as hybrid functionals.

Generalizations to Include Magnetic Fields

The DFT formalism described above breaks down, to various degrees, in the presence of a vector potential, i.e. a magnetic field. In such a situation, the one-to-one mapping between the ground-state electron density and wavefunction is lost. Generalizations to include the effects of magnetic fields have led to two different theories: current density functional theory (CDFT) and magnetic field density functional theory (BDFT). In both these theories, the functional used for the exchange and correlation must be generalized to include more than just the electron density. In current density functional theory, developed by Vignale and Rasolt, the functionals become dependent on both the electron density and the paramagnetic current density. In magnetic field density functional theory, developed by Salsbury, Grayce and Harris, the functionals depend on the electron density and the magnetic field, and the functional form can depend on the form of the magnetic field. In both of these theories it has been difficult to develop functionals beyond their equivalent to LDA, which are also readily implementable computationally. Recently an extension by Pan and Sahni extended the Hohenberg–Kohn theorem for varying magnetic fields using the density and the current density as fundamental variables.

Applications

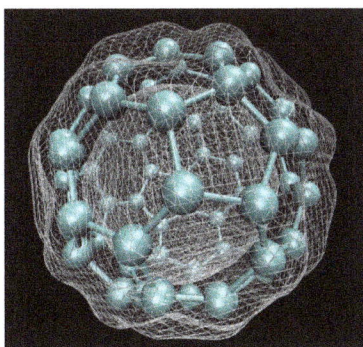

C_{60} with isosurface of ground-state electron density as calculated with DFT.

In general, density functional theory finds increasingly broad application in chemistry and materials science for the interpretation and prediction of complex system behavior at an atomic scale. Specifically, DFT computational methods are applied for synthesis-related systems and processing parameters. In such systems, experimental studies are often encumbered by inconsistent results and non-equilibrium conditions. Examples of contemporary DFT applications include studying the effects of dopants on phase transformation behavior in oxides, magnetic behavior in dilute magnetic semiconductor materials, and the study of magnetic and electronic behavior in ferroelectrics and dilute magnetic semiconductors. It has also been shown that DFT gives good results in the prediction of sensitivity of some nanostructures to environmental pollutants like sulfur dioxide or acrolein as well as prediction of mechanical properties.

In practice, Kohn–Sham theory can be applied in several distinct ways depending on what is being investigated. In solid state calculations, the local density approximations are still commonly used along with plane wave basis sets, as an electron gas approach is more appropriate for electrons which are delocalised through an infinite solid. In molecular calculations, however, more sophisticated functionals are needed, and a huge variety of exchange–correlation functionals have been developed for chemical applications. Some of these are inconsistent with the uniform electron gas approximation; however, they must reduce to LDA in the electron gas limit. Among physicists, one of the most widely used functionals is the revised Perdew–Burke–Ernzerhof exchange model (a direct generalized gradient parameterization of the free electron gas with no free parameters); however, this is not sufficiently calorimetrically accurate for gas-phase molecular calculations. In the chemistry community, one popular functional is known as BLYP (from the name Becke for the exchange part and Lee, Yang and Parr for the correlation part). Even more widely used is B3LYP, which is a hybrid functional in which the exchange energy, in this case from Becke's exchange functional, is combined with the exact energy from Hartree–Fock theory. Along with the component exchange and correlation functionals, three parameters define the hybrid functional, specifying how much of the exact exchange is mixed in. The adjustable parameters in hybrid functionals are generally fitted to a 'training set' of molecules. Although the results obtained with these functionals are usually sufficiently accurate for most applications, there is no systematic way of improving them (in contrast to some of the traditional wavefunction-based methods like configuration interaction or coupled cluster theory). In the current DFT approach it is not possible to estimate the error of the calculations without comparing them to other methods or experiments.

Thomas–Fermi Model

The predecessor to density functional theory was the Thomas–Fermi model, developed independently by both Thomas and Fermi in 1927. They used a statistical model to approximate the distribution of electrons in an atom. The mathematical basis postulated that electrons are distributed uniformly in phase space with two electrons in every h^3 of volume. For each element of coordinate space volume d^3r we can fill out a sphere of momentum space up to the Fermi momentum p_f:

$$\tfrac{4}{3}\pi p_f^3(\vec{r}).$$

Equating the number of electrons in coordinate space to that in phase space gives:

$$n(\vec{r}) = \frac{8\pi}{3h^3} p_f^3(\vec{r}).$$

Solving for p_f and substituting into the classical kinetic energy formula then leads directly to a kinetic energy represented as a functional of the electron density:

$$t_{TF}[n] = \frac{p^2}{2m_e} \propto \frac{\left(n^{\frac{1}{3}}\right)^2}{2m_e} \propto n^{\frac{2}{3}}(\vec{r}),$$

$$T_{TF}[n] = C_F \int n(\vec{r}) n^{\frac{2}{3}}(\vec{r}) \mathrm{d}^3 r = C_F \int n^{\frac{5}{3}}(\vec{r}) \mathrm{d}^3 r,$$

where,

$$C_F = \frac{3h^2}{10m_e} \left(\frac{3}{8\pi}\right)^{\frac{2}{3}}.$$

As such, they were able to calculate the energy of an atom using this kinetic energy functional combined with the classical expressions for the nucleus–electron and electron–electron interactions (which can both also be represented in terms of the electron density).

Although this was an important first step, the Thomas–Fermi equation's accuracy is limited because the resulting kinetic energy functional is only approximate, and because the method does not attempt to represent the exchange energy of an atom as a conclusion of the Pauli principle. An exchange energy functional was added by Dirac in 1928.

However, the Thomas–Fermi–Dirac theory remained rather inaccurate for most applications. The largest source of error was in the representation of the kinetic energy, followed by the errors in the exchange energy, and due to the complete neglect of electron correlation.

Teller (1962) showed that Thomas–Fermi theory cannot describe molecular bonding. This can be overcome by improving the kinetic energy functional.

The kinetic energy functional can be improved by adding the Weizsäcker correction:

$$T_W[n] = \frac{\hbar^2}{8m} \int \frac{|\nabla n(\vec{r})|^2}{n(\vec{r})} \mathrm{d}^3 r.$$

Hohenberg–Kohn Theorems

The Hohenberg–Kohn theorems relate to any system consisting of electrons moving under the influence of an external potential.

Theorem: The external potential (and hence the total energy), is a unique functional of the electron density.

If two systems of electrons, one trapped in a potential $v_1(\vec{r})$ and the other in $v_2(\vec{r})$, have the same ground-state density $n(\vec{r})$ then $v_1(\vec{r}) - v_2(\vec{r})$ is necessarily a constant.

Corollary: the ground state density uniquely determines the potential and thus all properties of the system, including the many-body wavefunction. In particular, the H–K functional, defined as F[n] = T[n] + U[n], is a universal functional of the density (not depending explicitly on the external potential).

Theorem: The functional that delivers the ground state energy of the system gives the lowest energy if and only if the input density is the true ground state density.

For any positive integer N and potential $\vec{v(r)}$, a density functional $F[n]$ exists such that.

$E_{(v,N)}[n] = F[n] + \int v(\vec{r}) n(\vec{r}) d^3 r$ obtains its minimal value at the ground-state density of N electrons in the potential $v(r) \rightarrow$. The minimal value of $E_{(v,N)}[n]$ is then the ground state energy of this system.

Pseudo-potentials

The many-electron Schrödinger equation can be very much simplified if electrons are divided in two groups: valence electrons and inner core electrons. The electrons in the inner shells are strongly bound and do not play a significant role in the chemical binding of atoms; they also partially screen the nucleus, thus forming with the nucleus an almost inert core. Binding properties are almost completely due to the valence electrons, especially in metals and semiconductors. This separation suggests that inner electrons can be ignored in a large number of cases, thereby reducing the atom to an ionic core that interacts with the valence electrons. The use of an effective interaction, a pseudopotential, that approximates the potential felt by the valence electrons, was first proposed by Fermi in 1934 and Hellmann in 1935. In spite of the simplification pseudo-potentials introduce in calculations, they remained forgotten until the late 1950s.

Ab Initio Pseudo-potentials

A crucial step toward more realistic pseudo-potentials was given by Topp and Hopfield and more recently Cronin, who suggested that the pseudo-potential should be adjusted such that they describe the valence charge density accurately. Based on that idea, modern pseudo-potentials are obtained inverting the free atom Schrödinger equation for a given reference electronic configuration and forcing the pseudo-wavefunctions to coincide with the true valence wave functions beyond a certain distance rl. The pseudo-wavefunctions are also forced to have the same norm as the true valence wavefunctions and can be written as:

$$R_l^{PP}(r) = R_{nl}^{AE}(r),$$

$$\int_0^{rl} \left| R_l^{PP}(r) \right|^2 r^2 \, dr = \int_0^{rl} \left| R_{nl}^{AE}(r) \right|^2 r^2 \, dr,$$

where $R_l(r)$ is the radial part of the wavefunction with angular momentum l; and PP and AE denote, respectively, the pseudo-wavefunction and the true (all-electron) wavefunction. The index n in the true wavefunctions denotes the valence level. The distance beyond which the true and the pseudo-wavefunctions are equal, rl, is also dependent on l.

Electron Smearing

The electrons of a system will occupy the lowest Kohn–Sham eigenstates up to a given energy level according to the Aufbau principle. This corresponds to the steplike Fermi–Dirac distribution at absolute zero. If there are several degenerate or close to degenerate eigenstates at the Fermi level, it is possible to get convergence problems, since very small perturbations may change the electron occupation. One way of damping these oscillations is to smear the electrons, i.e. allowing fractional occupancies. One approach of doing this is to assign a finite temperature to the electron Fermi–Dirac distribution. Other ways is to assign a cumulative Gaussian distribution of the electrons or using a Methfessel–Paxton method.

Frontier Molecular Orbital Theory

In chemistry, frontier molecular orbital theory is an application of MO theory describing HOMO/LUMO interactions.

In 1952, Kenichi Fukui published a paper in the Journal of Chemical Physics titled "A molecular theory of reactivity in aromatic hydrocarbons". Though widely criticized at the time, he later shared the Nobel Prize in Chemistry with Roald Hoffmann for his work on reaction mechanisms. Hoffman's work focused on creating a set of four pericyclic reactions in organic chemistry, based on orbital symmetry, which he coauthored with Robert Burns Woodward, entitled "The Conservation of Orbital Symmetry".

Fukui's own work looked at the frontier orbitals, and in particular the effects of the Highest Occupied Molecular Orbital (HOMO) and the Lowest Unoccupied Molecular Orbital (LUMO) on reaction mechanisms, which led to it being called Frontier Molecular Orbital Theory (FMO Theory). He used these interactions to better understand the conclusions of the Woodward–Hoffmann rules.

Theory

Fukui realized that a good approximation for reactivity could be found by looking at the frontier orbitals (HOMO/LUMO). This was based on three main observations of molecular orbital theory as two molecules interact:

- The occupied orbitals of different molecules repel each other.

- Positive charges of one molecule attract the negative charges of the other.

- The occupied orbitals of one molecule and the unoccupied orbitals of the other (especially the HOMO and LUMO) interact with each other causing attraction.

In general, the total energy change of the reactants on approach of the transition state is described by the Klopman-Salem equation, derived from perturbational MO theory. The first and second observations correspond to taking into consideration the filled-filled interaction and Coulombic interaction terms of the equation, respectively. With respect to the third observation, primary consideration of the HOMO-LUMO interaction is justified by the fact that the largest contribution in the filled-unfilled interaction term of the Klopman-Salem equation comes from molecular orbitals r and s that are closest in energy (i.e., smallest $E_r - E_s$ value). From these observations, frontier molecular orbital (FMO) theory simplifies prediction of reactivity to analysis of the interaction between the more energetically matched HOMO-LUMO pairing of the two reactants. In addition to providing a unified explanation of diverse aspects of chemical reactivity and selectivity, it agrees with the predictions of the Woodward–Hoffmann orbital symmetry and Dewar-Zimmerman aromatic transition state treatments of thermal pericyclic reactions, which are summarized in the following selection rule:

"A ground-state pericyclic change is symmetry-allowed when the total number of $(4q+2)_s$ and $(4r)_a$ components is odd".

$(4q+2)_s$ refers to the number of aromatic, suprafacial electron systems; likewise, $(4r)_a$ refers to antiaromatic, antarafacial systems. It can be shown that if the total number of these systems is odd then the reaction is thermally allowed.

Applications

Cycloadditions

A cycloaddition is a reaction that simultaneously forms at least two new bonds, and in doing so, converts two or more open-chain molecules into rings. The transition states for these reactions typically involves the electrons of the molecules moving in continuous rings, making it a pericyclic reaction. These reactions can be predicted by the Woodward–Hoffmann rules and thus are closely approximated by FMO Theory.

The Diels–Alder reaction between maleic anhydride and cyclopentadiene is allowed by the Woodward–Hoffmann rules because there are six electrons moving suprafacially and no electrons moving antarafacially. Thus, there is one $(4q + 2)_s$ component and no $(4r)_a$ component, which means the reaction is allowed thermally.

FMO theory also finds that this reaction is allowed and goes even further by predicting its stereoselectivity, which is unknown under the Woodward-Hoffmann rules. Since this is a [4 + 2], the reaction can be simplified by considering the reaction between butadiene and ethene. The HOMO of butadiene and the LUMO of ethene are both antisymmetric (rotationally symmetric), meaning the reaction is allowed.

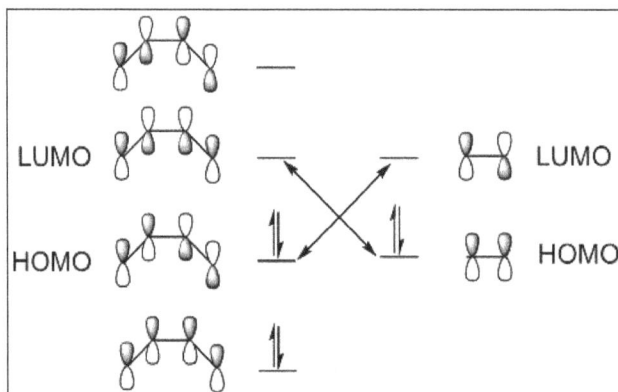

In terms of the stereoselectivity of the reaction between maleic anhydride and cyclopentadiene, the endo-product is favored, a result best explained through FMO theory. The maleic anhydride is an electron-withdrawing species that makes the dieneophile electron deficient, forcing the regular Diels–Alder reaction. Thus, only the reaction between the HOMO of cyclopentadiene and the

LUMO of maleic anhydride is allowed. Furthermore, though the exo-product is the more thermo-dynamically stable isomer, there are secondary (non-bonding) orbital interactions in the endo-transition state, lowering its energy and making the reaction towards the endo- product faster, and therefore more kinetically favorable. Since the exo-product has primary (bonding) orbital interactions it can still form, but since the endo-product forms faster it is the major product.

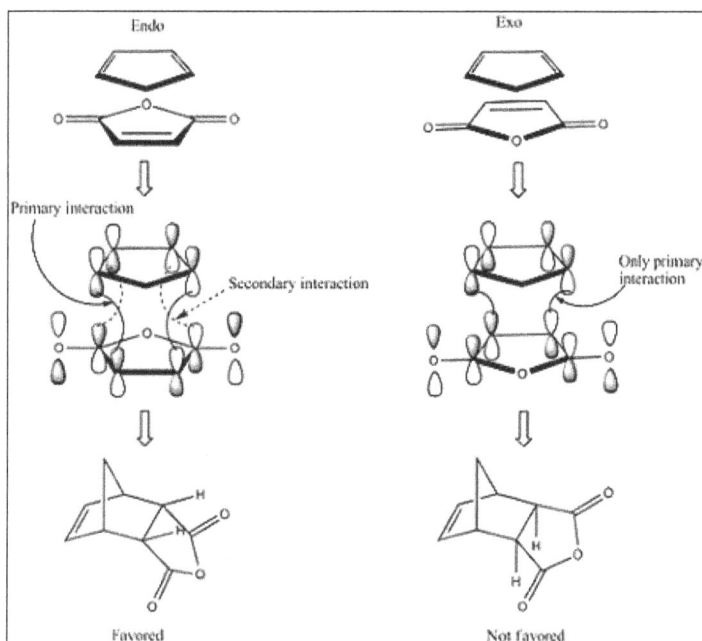

*The HOMO of ethene and the LUMO of butadiene are both symmetric, meaning the reaction between these species is allowed as well. This is referred to as the "inverse electron demand Diels–Alder".

Sigmatropic Reactions

A sigmatropic rearrangement is a reaction in which a sigma bond moves across a conjugated pi system with a concomitant shift in the pi bonds. The shift in the sigma bond may be antarafacial or suprafacial. In the example of a [1,5] shift in pentadiene, if there is a suprafacial shift, there is 6 e$^-$ moving suprafacially and none moving antarafacially, implying this reaction is allowed by the Woodward–Hoffmann rules. For an antarafacial shift, the reaction is not allowed.

These results can be predicted with FMO theory by observing the interaction between the HOMO and LUMO of the species. To use FMO theory, the reaction should be considered as two separate ideas: (1) whether or not the reaction is allowed, and (2) which mechanism the reaction proceeds though. In the case of a [1,5] shift on pentadiene, the HOMO of the sigma bond (i.e. a constructive bond) and the LUMO of butadiene on the remaining 4 carbons is observed. Assuming the reaction happens suprafacially, the shift results with the HOMO of butadiene on the 4 carbons that are not involved in the sigma bond of the product. Since the pi system changed from the LUMO to the HOMO, this reaction is allowed (though it would not be allowed if the pi system went from LUMO to LUMO).

To explain why the reaction happens suprafacially, first notice that the terminal orbitals are in the same phase. For there to be a constructive sigma bond formed after the shift, the reaction would have to be suprafacial. If the species shifted antarafacially then it would form an antibonding orbital and there would not be a constructive sigma shift.

It is worth noting that in propene the shift would have to be antarafacial, but since the molecule is very small that twist is not possible and the reaction is not allowed.

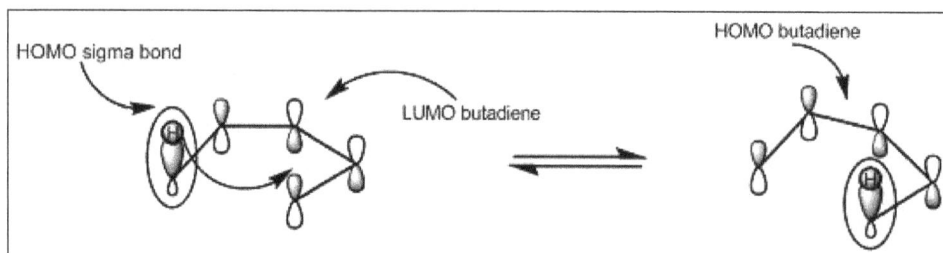

Electrocyclic Reactions

An electrocyclic reaction is a pericyclic reaction involving the net loss of a pi bond and creation of a sigma bond with formation of a ring. This reaction proceeds through either a conrotatory or disrotatory mechanism. In the conrotatory ring opening of cyclobutene, there are two electrons moving suprafacially (on the pi bond) and two moving antarafacially (on the sigma bond). This means there is one $4q + 2$ suprafacial system and no $4r$ antarafacial system; thus the conrotatory process is thermally allowed by the Woodward–Hoffmann rules.

The HOMO of the sigma bond (i.e. a constructive bond) and the LUMO of the pi bond are important in the FMO theory consideration. If the ring opening uses a conrotatory process then the reaction results with the HOMO of butadiene. As in the previous examples the pi system moves from a LUMO species to a HOMO species, meaning this reaction is allowed.

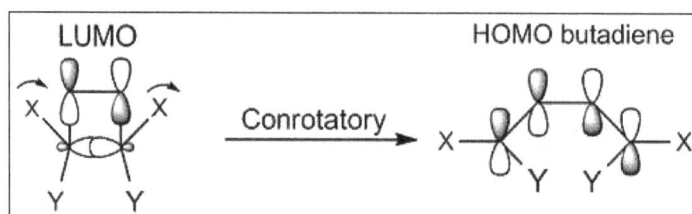

Time-dependent Density Functional Theory

Time-dependent density functional theory (TDDFT) is a quantum mechanical theory used in physics and chemistry to investigate the properties and dynamics of many-body systems in the presence of time-dependent potentials, such as electric or magnetic fields. The effect of such fields on molecules and solids can be studied with TDDFT to extract features like excitation energies, frequency-dependent response properties, and photoabsorption spectra.

TDDFT is an extension of density functional theory (DFT), and the conceptual and computational foundations are analogous – to show that the (time-dependent) wave function is equivalent to the (time-dependent) electronic density, and then to derive the effective potential of a fictitious non-interacting system which returns the same density as any given interacting system. The issue of constructing such a system is more complex for TDDFT, most notably because the time-dependent effective potential at any given instant depends on the value of the density at all previous times. Consequently, the development of time-dependent approximations for the implementation of TDDFT is behind that of DFT, with applications routinely ignoring this memory requirement.

The formal foundation of TDDFT is the Runge-Gross (RG) theorem – the time-dependent analogue of the Hohenberg-Kohn (HK) theorem. The RG theorem shows that, for a given initial wavefunction, there is a unique mapping between the time-dependent external potential of a system and its time-dependent density. This implies that the many-body wavefunction, depending upon 3N variables, is equivalent to the density, which depends upon only 3, and that all properties of a system can thus be determined from knowledge of the density alone. Unlike in DFT, there is no general minimization principle in time-dependent quantum mechanics. Consequently, the proof of the RG theorem is more involved than the HK theorem.

Given the RG theorem, the next step in developing a computationally useful method is to determine the fictitious non-interacting system which has the same density as the physical (interacting) system of interest. As in DFT, this is called the (time-dependent) Kohn-Sham system. This system is formally found as the stationary point of an action functional defined in the Keldysh formalism.

The most popular application of TDDFT is in the calculation of the energies of excited states of isolated systems and, less commonly, solids. Such calculations are based on the fact that the linear response function – that is, how the electron density changes when the external potential changes – has poles at the exact excitation energies of a system. Such calculations require, in addition to the exchange-correlation potential, the exchange-correlation kernel – the functional derivative of the exchange-correlation potential with respect to the density.

Formalism

Runge-Gross Theorem

The approach of Runge and Gross considers a single-component system in the presence of a time-dependent scalar field for which the Hamiltonian takes the form:

$$\hat{H}(t) = \hat{T} + \hat{V}_{ext}(t) + \hat{W},$$

where T is the kinetic energy operator, W the electron-electron interaction, and $V_{ext}(t)$ the external potential which along with the number of electrons defines the system. Nominally, the external potential contains the electrons' interaction with the nuclei of the system. For non-trivial time-dependence, an additional explicitly time-dependent potential is present which can arise, for example, from a time-dependent electric or magnetic field. The many-body wavefunction evolves according to the time-dependent Schrödinger equation under a single initial condition,

$$\hat{H}(t) \, | \, \Psi(t) \rangle = i\hbar \frac{\partial}{\partial t} \, | \, \Psi(t) \rangle, \quad | \, \Psi(0) \rangle = | \, \Psi \rangle.$$

Employing the Schrödinger equation as its starting point, the Runge-Gross theorem shows that at any time, the density uniquely determines the external potential. This is done in two steps:

- Assuming that the external potential can be expanded in a Taylor series about a given time, it is shown that two external potentials differing by more than an additive constant generate different current densities.

- Employing the continuity equation, it is then shown that for finite systems, different current densities correspond to different electron densities.

Time-Dependent Kohn-Sham System

For a given interaction potential, the RG theorem shows that the external potential uniquely determines the density. The Kohn-Sham approaches chooses a non-interacting system (that for which the interaction potential is zero) in which to form the density that is equal to the interacting system. The advantage of doing so lies in the ease in which non-interacting systems can be solved – the wave function of a non-interacting system can be represented as a Slater determinant of single-particle orbitals, each of which are determined by a single partial differential equation in three variable – and that the kinetic energy of a non-interacting system can be expressed exactly in terms of those orbitals. The problem is thus to determine a potential, denoted as $v_s(\mathbf{r},t)$ or $v_{KS}(\mathbf{r},t)$, that determines a non-interacting Hamiltonian, H_s,

$$\hat{H}_s(t) = \hat{T} + \hat{V}_s(t),$$

which in turn determines a determinantal wave function:

$$\hat{H}_s(t)\,|\,\Phi(t)\rangle = i\frac{\partial}{\partial t}\,|\,\Phi(t)\rangle, \quad |\,\Phi(0)\rangle = |\,\Phi\rangle,$$

which is constructed in terms of a set of N orbitals which obey the equation,

$$\left(-\frac{1}{2}\nabla^2 + v_s(\mathbf{r},t)\right)\phi_i(\mathbf{r},t) = i\frac{\partial}{\partial t}\phi_i(\mathbf{r},t) \quad \phi_i(\mathbf{r},0) = \phi_i(\mathbf{r}),$$

and generate a time-dependent density:

$$\rho_s(\mathbf{r},t) = \sum_{i=1}^{N_b} f_i(t)\,|\,\phi_i(\mathbf{r},t)\,|^2,$$

such that ρ_s is equal to the density of the interacting system at all times:

$$\rho_s(\mathbf{r},t) = \rho(\mathbf{r},t).$$

Note that in the expression of density above, the summation is over *all* N_b Kohn-Sham orbitals and is the time-dependent occupation number for orbital $f_i(t)$. If the potential $v_s(\mathbf{r},t)$ can be determined, or at the least well-approximated, then the original Schrödinger equation, a single partial differential equation in $3N$ variables, has been replaced by N differential equations in 3 dimensions, each differing only in the initial condition.

The problem of determining approximations to the Kohn-Sham potential is challenging. Analogously to DFT, the time-dependent KS potential is decomposed to extract the external potential of the system and the time-dependent Coulomb interaction, v_J. The remaining component is the exchange-correlation potential:

$$v_s(\mathbf{r},t) = v_{ext}(\mathbf{r},t) + v_J(\mathbf{r},t) + v_{xc}(\mathbf{r},t).$$

In their seminal paper, Runge and Gross approached the definition of the KS potential through an action-based argument starting from the Dirac action:

$$A[\Psi] = \int dt \, \langle \Psi(t) \mid H - i\frac{\partial}{\partial t} \mid \Psi(t) \rangle.$$

Treated as a functional of the wave function, $A[\Psi]$, variations of the wave function yield the many-body Schrödinger equation as the stationary point. Given the unique mapping between densities and wave function, Runge and Gross then treated the Dirac action as a density functional,

$$A[\rho] = A[\Psi[\rho]],$$

and derived a formal expression for the exchange-correlation component of the action, which determines the exchange-correlation potential by functional differentiation. Later it was observed that an approach based on the Dirac action yields paradoxical conclusions when considering the causality of the response functions it generates. The density response function, the functional derivative of the density with respect to the external potential, should be causal: a change in the potential at a given time can not affect the density at earlier times. The response functions from the Dirac action however are symmetric in time so lack the required causal structure. An approach which does not suffer from this issue was later introduced through an action based on the Keldysh formalism of complex-time path integration. An alternative resolution of the causality paradox through a refinement of the action principle in real time has been recently proposed by Vignale.

Linear Response TDDFT

Linear-response TDDFT can be used if the external perturbation is small in the sense that it does not completely destroy the ground-state structure of the system. In this case one can analyze the linear response of the system. This is a great advantage as, to first order, the variation of the system will depend only on the ground-state wave-function so that we can simply use all the properties of DFT.

Consider a small time-dependent external perturbation $\delta V^{ext}(t)$. This gives:

$$H'(t) = H + \delta V^{ext}(t)$$

$$H'_{KS}[\rho](t) = H_{KS}[\rho] + \delta V_H[\rho](t) + \delta V_{xc}[\rho](t) + \delta V^{ext}(t)$$

and looking at the linear response of the density:

$$\delta\rho(\mathbf{r}t) = \chi(\mathbf{r}t,\mathbf{r}'t')\delta V^{ext}(\mathbf{r}'t')$$

$$\delta\rho(\mathbf{r}t) = \chi_{KS}(\mathbf{r}t,\mathbf{r}'t')\delta V^{eff}[\rho](\mathbf{r}'t')$$

where $\delta V^{eff}[\rho](t) = \delta V^{ext}(t) + \delta V_H[\rho](t) + \delta V_{xc}[\rho](t)$ Here and in the following it is assumed that primed variables are integrated.

Within the linear-response domain, the variation of the Hartree (H) and the exchange-correlation (xc) potential to linear order may be expanded with respect to the density variation:

$$\delta V_H[\rho](\mathbf{r}) = \frac{\delta V_H[\rho]}{\delta \rho} \delta \rho = \frac{1}{|\mathbf{r} - \mathbf{r}'|} \delta \rho(\mathbf{r}')$$

and

$$\delta V_{xc}[\rho](\mathbf{r}) = \frac{\delta V_{xc}[\rho]}{\delta \rho} \delta \rho = f_{xc}(\mathbf{r}t, \mathbf{r}'t') \delta \rho(\mathbf{r}')$$

Finally, inserting this relation in the response equation for the KS system and comparing the resultant equation with the response equation for the physical system yields the Dyson equation of TDDFT:

$$\chi(\mathbf{r}_1 t_1, \mathbf{r}_2 t_2) = \chi_{KS}(\mathbf{r}_1 t_1, \mathbf{r}_2 t_2) + \chi_{KS}(\mathbf{r}_1 t_1, \mathbf{r}'_2 t'_2) \left(\frac{1}{|\mathbf{r}'_2 - \mathbf{r}'_1|} + f_{xc}(\mathbf{r}'_2 t'_2, \mathbf{r}'_1 t'_1) \right) \chi(\mathbf{r}'_1 t'_1, \mathbf{r}_2 t_2)$$

From this last equation it is possible to derive the excitation energies of the system, as these are simply the poles of the response function.

Other linear-response approaches include the Casida formalism (an expansion in electron-hole pairs) and the Sternheimer equation (density-functional perturbation theory).

References

- 8-1-valence-bond-theory, chapter, chemistry: opentextbc.ca, Retrieved 16 March, 2019

- Baran, E. (2000). "Mean amplitudes of vibration of the pentagonal pyramidal xeof–5 and IOF2–5 anions". J. Fluorine Chem. 101: 61–63. Doi:10.1016/S0022-1139(99)00194-3

- Topicreview, genchem: chemed.chem.purdue.edu, Retrieved 17 April, 2019

- Petrucci, R. H.; W. S., Harwood; F. G., Herring (2002). General Chemistry: Principles and Modern Applications (8th ed.). Prentice-Hall. Pp. 413–414 (Table 11.1). ISBN 978-0-13-014329-7

- Zimmerli, Urs; Parrinello, Michele; Koumoutsakos, Petros (2004). "Dispersion corrections to density functionals for water aromatic interactions". Journal of Chemical Physics. 120 (6): 2693–2699. Bibcode:-2004jchph.120.2693Z. Doi:10.1063/1.1637034. PMID 15268413

6

Qualitative Theories of Chemical Bonding

Qualitative theory of chemical bonding includes atomic orbitals, molecular orbitals, electronic configuration, Huckel approximation and Huckel molecular orbital theory. The topics elaborated in this chapter will help in gaining a better perspective about these qualitative theories of chemical bonding.

Atomic Orbital

When a planet moves around the sun, its definite path, called an orbit, can be plotted. A drastically simplified view of the atom looks similar, in which the electrons orbit around the nucleus. The truth is different; electrons, in fact, inhabit regions of space known as *orbitals*. Orbits and orbitals sound similar, but they have quite different meanings. It is essential to understand the difference between them.

The Impossibility of Drawing Orbits for Electrons

To plot a path for something, the exact location and trajectory of the object must be known. This is imossible for electrons. The Heisenberg Uncertainty Principle states that it is impossible to define with absolute precision, at the same time, both the position and the momentum of an electron. That makes it impossible to plot an orbit for an electron around a nucleus.

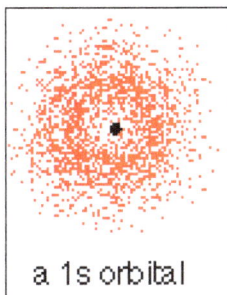

a 1s orbital

Hydrogen's electron - the 1s orbital.

Consider a single hydrogen atom: at a particular instant, the position of the electron is plotted. The position is plotted again soon afterward, and it is in a different position. There is no way to tell how it moved from the first place to the second. This process is repeated many times, eventually creating a 3D map of the places that the electron is likely to be found.

In the hydrogen case, the electron can be found anywhere within a spherical space surrounding the nucleus. The figure above shows a cross-sectionof this spherical space. 95% of the time (or any arbitrary, high percentage), the electron is found within a fairly easily defined region of space quite close to the nucleus. Such a region of space is called an orbital, and it can be thought of as the region of space the electron inhabits. It is impossible to know what the electron is doing inside the orbital, so the electron's actions are ignored completely. All that can be said is that if an electron is in a particular orbital, it has a particular, definable energy.

Each Orbital has a Name

The orbital occupied by the hydrogen electron is called a 1s orbital. The number "1" represents the fact that the orbital is in the energy level closest to the nucleus. The letter "s" indicates the shape of the orbital: s orbitals are spherically symmetric around the nucleus-they look like hollow balls made of chunky material with the nucleus at the center.

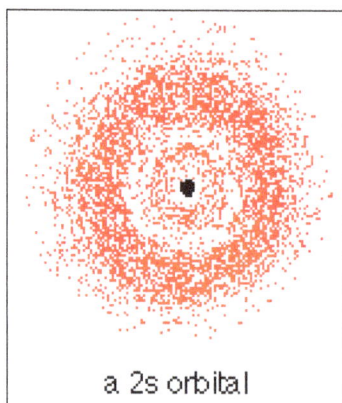

a 2s orbital

Hydrogen's electron - the 2s orbital.

The orbital shown above is a 2s orbital. This is similar to a 1s orbital, except that the region where there is the greatest chance of finding the electron is further from the nucleus. This is an orbital at the second energy level. There is another region of slightly higher electron density (where the dots are thicker) nearer the nucleus ("electron density" is another way of describing the likelihood of an electron at a particular place).

2s (and 3s, 4s, etc). electrons spend some of their time closer to the nucleus than might be expected. The effect of this is to slightly reduce the energy of electrons in s orbitals. The nearer the nucleus the electrons get, the lower their energy. 3s, 4s (etc). orbitals are progressively further from the nucleus.

P Orbitals

Not all electrons inhabit s orbitals (in fact, very few electrons live occupy s orbitals). At the first energy level, the only orbital available to electrons is the 1s orbital, but at the second level, as well as a 2s orbital, there are 2p orbitals. A p orbital is shaped like 2 identical balloons tied together at the nucleus. The orbital shows where there is a 95% chance of finding a particular electron.

Imagine a horizontal plane through the nucleus, with one lobe of the orbital above the plane and the other beneath it; there is a zero probability of finding the electron on that plane. How does the electron get from one lobe to the other if it can never pass through the plane of the nucleus? At an introductory level, it must simply be accepted. To find out more, read about the wave nature of electrons.

At any one energy level it is possible to have three absolutely equivalent p orbitals pointing mutually at right angles to each other. These are arbitrarily given the symbols p_x, p_y and p_z. This is simply for convenience; the x, y, and z directions change constantly as the atom tumbles in space.

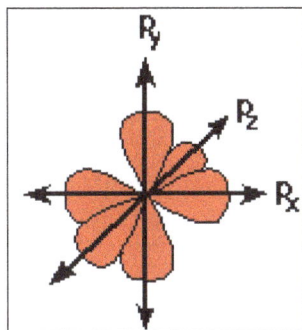

Hydrogen's electron - the 2p orbitals.

The p orbitals at the second energy level are called $2p_x$, $2p_y$ and $2p_z$. There are similar orbitals at subsequent levels: $3p_x$, $3p_y$, $3p_z$, $4p_x$, $4p_y$, $4p_z$, and so on. All levels except for the first level have p orbitals. At the higher levels the lobes are more elongated, with the most likely place to find the electron more distant from the nucleus.

d and f Orbitals

In addition to s and p orbitals, there are two other sets of orbitals that become available for electrons to inhabit at higher energy levels. At the third level, there is a set of five d orbitals (with complicated shapes and names) as well as the 3s and 3p orbitals ($3p_x$, $3p_y$, $3p_z$). At the third level there are nine total orbitals. At the fourth level, as well the 4s and 4p and 4d orbitals there are an additional seven f orbitals, adding up to 16 orbitals in all. s, p, d and f orbitals are then available at all higher energy levels as well.

Fitting Electrons into Orbitals

An atom can be pictured as a very strange house (somewhat an inverted pyramid), with the nucleus living on the ground floor, and then various rooms (orbitals) on the higher floors occupied by the electrons. On the first floor there is only 1 room (the 1s orbital); on the second floor there are 4 rooms (the 2s, $2p_x$, $2p_y$ and $2p_z$ orbitals); on the third floor there are 9 rooms (one 3s orbital, three 3p orbitals and five 3d orbitals); and so on. However, the rooms are not large: each orbital can only hold 2 electrons.

Electrons in Boxes

A convenient way of showing the orbitals that the electrons live in is to draw "electrons-in-boxes". Orbitals can be represented as boxes with the electrons depicted with arrows. Often an up-arrow and a down-arrow are used to show that the electrons are different. The need for all electrons in an atom to be different originates from quantum theory. If the electrons inhabit different orbitals, they can have identical properties, but if they are both in the same orbital there must be a distinction between them. Quantum theory allocates them a property known as "spin," represented by the direction the arrow is pointing.

A 1s orbital holding 2 electrons is drawn as shown on the right, but it can be written even more quickly as 1s². This is read as "one s two," not as "one s squared". Do not confuse the energy level with the number of electrons in this notation.

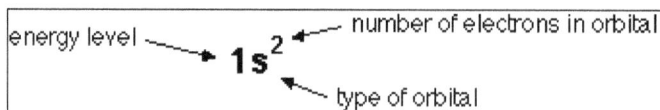

Molecular Orbitals

A molecular orbital is an orbital or wave function of a molecule's electron. The function may be used to calculate the probability of finding an electron within a specified space or to predict the molecule's chemical and physical properties. Robert Mulliken introduced the term "orbital" in 1932 to describe a one-electron orbital wave function.

Electrons around a molecule can be associated with more than one atom and are often expressed as a combination of atomic orbitals. Atomic orbitals within a molecule can interact if they have compatible symmetries. The number of molecular orbitals is equal to the number of atomic orbitals combined to form a molecule.

Molecular Orbital Diagram

A molecular orbital diagram, or MO diagram, is a qualitative descriptive tool explaining chemical bonding in molecules in terms of molecular orbital theory in general and the linear combination of atomic orbitals (LCAO) method in particular. A fundamental principle of these theories is that as atoms bond to form molecules, a certain number of atomic orbitals combine to form the same number of molecular orbitals, although the electrons involved may be redistributed among the orbitals. This tool is very well suited for simple diatomic molecules such as dihydrogen, dioxygen, and carbon monoxide but becomes more complex when discussing even comparatively simple polyatomic molecules, such as methane. MO diagrams can explain why some molecules exist and others do not. They can also predict bond strength, as well as the electronic transitions that can take place.

Basics

Molecular orbital diagrams are diagrams of molecular orbital (MO) energy levels, shown as short horizontal lines in the center, flanked by constituent atomic orbital (AO) energy levels for comparison, with the energy levels increasing from the bottom to the top. Lines, often dashed diagonal lines, connect MO levels with their constituent AO levels. Degenerate energy levels are commonly shown side by side. Appropriate AO and MO levels are filled with electrons by the Pauli Exclusion Principle, symbolized by small vertical arrows whose directions indicate the electron spins. The AO or MO shapes themselves are often not shown on these diagrams. For a diatomic molecule, an MO diagram effectively shows the energetics of the bond between the two atoms, whose AO unbonded energies are shown on the sides. For simple polyatomic molecules with a "central atom" such as methane (CH_4) or carbon dioxide (CO_2), a MO diagram may show one of the identical bonds to the central atom. For other polyatomic molecules, an MO diagram may show one or more bonds of interest in the molecules, leaving others out for simplicity. Often even for simple molecules, AO and MO levels of inner orbitals and their electrons may be omitted from a diagram for simplicity.

In MO theory molecular orbitals form by the overlap of atomic orbitals. Because σ bonds feature greater overlap than π bonds, σ bonding and σ* antibonding orbitals feature greater energy splitting (separation) than π and π* orbitals. The atomic orbital energy correlates with electronegativity as more electronegative atoms hold their electrons more tightly, lowering their energies. Sharing of molecular orbitals between atoms is more important when the atomic orbitals have comparable energy; when the energies differ greatly the orbitals tend to be localized on one atom and the mode of bonding becomes ionic. A second condition for overlapping atomic orbitals is that they have the same symmetry.

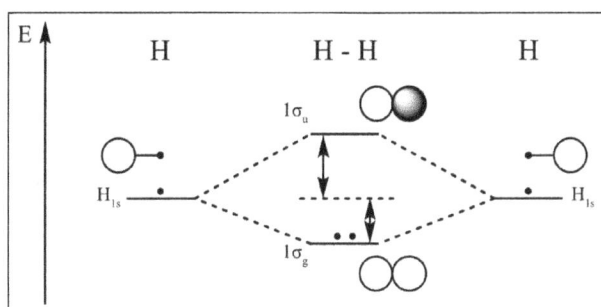

MO diagram for dihydrogen. Here electrons are shown by dots.

Two atomic orbitals can overlap in two ways depending on their phase relationship (or relative signs for real orbitals). The phase (or sign) of an orbital is a direct consequence of the wave-like properties of electrons. In graphical representations of orbitals, orbital sign is depicted either by a plus or minus sign (which has no relationship to electric charge) or by shading one lobe. The sign of the phase itself does not have physical meaning except when mixing orbitals to form molecular orbitals.

Two same-sign orbitals have a constructive overlap forming a molecular orbital with the bulk of the electron density located between the two nuclei. This MO is called the bonding orbital and its energy is lower than that of the original atomic orbitals. A bond involving molecular orbitals which are symmetric with respect to any rotation around the bond axis is called a sigma bond (σ-bond). If the phase cycles once while rotating round the axis, the bond is a pi bond (π-bond). Symmetry labels are further defined by whether the orbital maintains its original character after an inversion about its center; if it does, it is defined gerade, g. If the orbital does not maintain its original character, it is ungerade, u.

Atomic orbitals can also interact with each other out-of-phase which leads to destructive cancellation and no electron density between the two nuclei at the so-called nodal plane depicted as a perpendicular dashed line. In this anti-bonding MO with energy much higher than the original AO's, any electrons present are located in lobes pointing away from the central internuclear axis. For a corresponding σ-bonding orbital, such an orbital would be symmetrical but differentiated from it by an asterisk as in σ^*. For a π-bond, corresponding bonding and antibonding orbitals would not have such symmetry around the bond axis and be designated π and π^*, respectively.

The next step in constructing an MO diagram is filling the newly formed molecular orbitals with electrons. Three general rules apply:

- The Aufbau principle states that orbitals are filled starting with the lowest energy.

- The Pauli exclusion principle states that the maximum number of electrons occupying an orbital is two, with opposite spins.

- Hund's rule states that when there are several MO's with equal energy, the electrons occupy the MO's one at a time before two electrons occupy the same MO.

The filled MO highest in energy is called the Highest Occupied Molecular Orbital or HOMO and the empty MO just above it is then the Lowest Unoccupied Molecular Orbital or LUMO. The electrons in the bonding MO's are called bonding electrons and any electrons in the antibonding orbital would be called antibonding electrons. The reduction in energy of these electrons is the driving force for chemical bond formation. Whenever mixing for an atomic orbital is not possible for reasons of symmetry or energy, a non-bonding MO is created, which is often quite similar to and has energy level equal or close to its constituent AO, thus not contributing to bonding energetics. The resulting electron configuration can be described in terms of bond type, parity and occupancy for example dihydrogen $1\sigma_g^2$. Alternatively it can be written as a molecular term symbol e.g. $^1\Sigma_g^+$ for dihydrogen. Sometimes, the letter **n** is used to designate a non-bonding orbital.

For a stable bond, the bond order, defined as:

$$\text{Bond Order} = \frac{(\text{No. of electrons in bonding MOs}) - (\text{No. of electrons in anti-bonding MOs})}{2}$$

must be positive.

The relative order in MO energies and occupancy corresponds with electronic transitions found in photoelectron spectroscopy (PES). In this way it is possible to experimentally verify MO theory. In general, sharp PES transitions indicate nonbonding electrons and broad bands are indicative of bonding and antibonding delocalized electrons. Bands can resolve into fine structure with spacings corresponding to vibrational modes of the molecular cation (see Franck–Condon principle). PES energies are different from ionisation energies which relates to the energy required to strip off the nth electron after the first n – 1 electrons have been removed. MO diagrams with energy values can be obtained mathematically using the Hartree–Fock method. The starting point for any MO diagram is a predefined molecular geometry for the molecule in question. An exact relationship between geometry and orbital energies is given in Walsh diagrams.

s-p Mixing

The phenomenon of s-p mixing occurs when molecular orbitals of the same symmetry formed from the combination of 2s and 2p atomic orbitals are close enough in energy to further interact, which can lead to a change in the expected order of orbital energies. When molecular orbitals are formed, they are mathematically obtained from linear combinations of the starting atomic orbitals. Generally, in order to predict their relative energies, it is sufficient to consider only one atomic orbital from each atom to form a pair of molecular orbitals, as the contributions from the others are negligible. For instance, in dioxygen the $3\sigma_g$ MO can be roughly considered to be formed from interaction of oxygen $2p_z$ AOs only. It is found to be lower in energy than the $1\pi_u$ MO, both experimentally and from more sophisticated computational models, so that the expected order of filling is the $3\sigma_g$ before the $1\pi_u$. Hence the approximation to ignore the effects of further interactions is valid. However, experimental and computational results for homonuclear diatomics from Li_2 to N_2 and certain heteronuclear combinations such as CO and NO show that the $3\sigma_g$ MO is higher in energy than (and therefore filled after) the $1\pi_u$ MO. This can be rationalised as the first-approximation $3\sigma_g$ has a suitable symmetry to interact with the $2\sigma_g$ bonding MO formed from the 2s AOs. As a result, the $2\sigma_g$ is lowered in energy, whilst the $3\sigma_g$ is raised. For the aforementioned molecules this results in the $3\sigma_g$ being higher in energy than the $1\pi_u$ MO, which is where s-p mixing is most evident. Likewise, interaction between the $2\sigma_u^*$ and $3\sigma_u^*$ MOs leads to a lowering in energy of the former and a raising in energy of the latter. However this is of less significance than the interaction of the bonding MOs.

Diatomic MO Diagrams

A diatomic molecular orbital diagram is used to understand the bonding of a diatomic molecule. MO diagrams can be used to deduce magnetic properties of a molecule and how they change with ionization. They also give insight to the bond order of the molecule, how many bonds are shared between the two atoms.

The energies of the electrons are further understood by applying the Schrödinger equation to a molecule. Quantum Mechanics is able to describe the energies exactly for single electron systems but can be approximated precisely for multiple electron systems using the Born-Oppenheimer Approximation, such that the nuclei are assumed stationary. The LCAO-MO method is used in conjunction to further describe the state of the molecule.

Diatomic molecules consist of a bond between only two atoms. They can be broken into two categories: homonuclear and heteronuclear. A homonuclear diatomic molecule is one composed of two atoms of the same element. Examples are H_2, O_2, and N_2. A heteronuclear diatomic molecule is composed of two atoms of two different elements. Examples include CO, HCl, and NO.

Dihydrogen

The smallest molecule, hydrogen gas exists as dihydrogen (H-H) with a single covalent bond between two hydrogen atoms. As each hydrogen atom has a single 1s atomic orbital for its electron, the bond forms by overlap of these two atomic orbitals. In the figure the two atomic orbitals are depicted on the left and on the right. The vertical axis always represents the orbital energies. Each atomic orbital is singly occupied with an up or down arrow representing an electron.

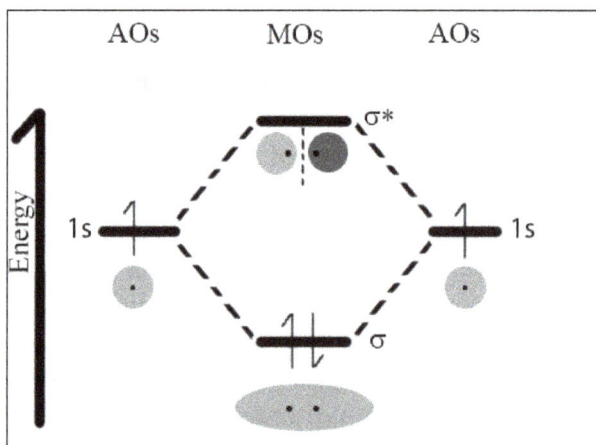

H2 Molecular Orbital Diagram.

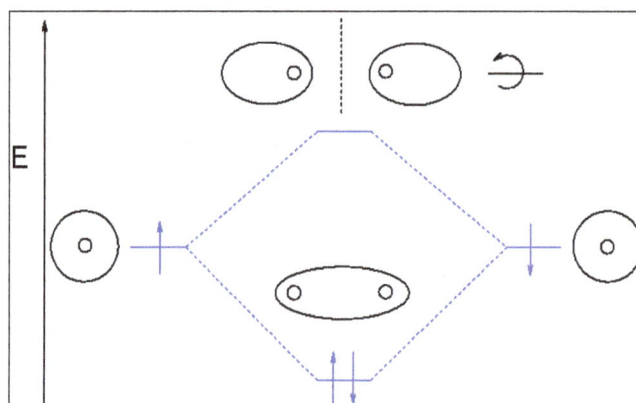

MO diagram of dihydrogen.

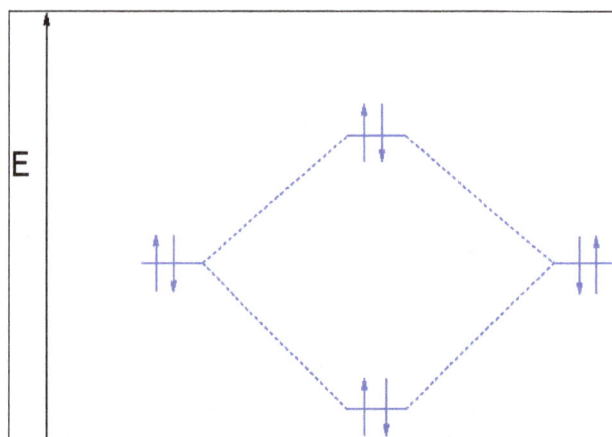

Bond breaking in MO diagram.

Application of MO theory for dihydrogen results in having both electrons in the bonding MO with electron configuration $1\sigma_g^2$. The bond order for dihydrogen is $(2-0)/2 = 1$. The photoelectron spectrum of dihydrogen shows a single set of multiplets between 16 and 18 eV (electron volts).

The dihydrogen MO diagram helps explain how a bond breaks. When applying energy to dihydrogen, a molecular electronic transition takes place when one electron in the bonding MO is promoted to the antibonding MO. The result is that there is no longer a net gain in energy.

The superposition of the two 1s atomic orbitals leads to the formation of the σ and σ* molecular orbitals. Two atomic orbitals in phase create a larger electron density, which leads to the σ orbital. If the two 1s orbitals are not in phase, a node between them causes a jump in energy, the σ* orbital. From the diagram you can deduce the bond order, how many bonds are formed between the two atoms. For this molecule it is equal to one. Bond order can also give insight to how close or stretched a bond has become if a molecule is ionized.

Dihelium and Diberyllium

Dihelium (He-He) is a hypothetical molecule and MO theory helps to explain why dihelium does not exist in nature. The MO diagram for dihelium looks very similar to that of dihydrogen, but each helium has two electrons in its 1s atomic orbital rather than one for hydrogen, so there are now four electrons to place in the newly formed molecular orbitals.

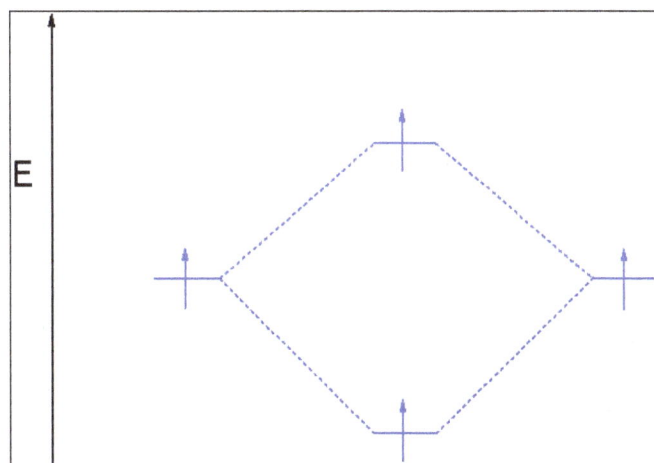

MO diagram of dihelium.

The only way to accomplish this is by occupying both the bonding and antibonding orbitals with two electrons, which reduces the bond order $((2-2)/2)$ to zero and cancels the net energy stabilization. However, by removing one electron from dihelium, the stable gas-phase species He_2^+ ion is formed with bond order $1/2$.

Another molecule that is precluded based on this principle is diberyllium. Beryllium has an electron configuration $1s^2 2s^2$, so there are again two electrons in the valence level. However, the 2s can mix with the 2p orbitals in diberyllium, whereas there are no p orbitals in the valence level of hydrogen or helium. This mixing makes the antibonding $1\sigma_u$ orbital slightly less antibonding than the bonding $1\sigma_g$ orbital is bonding, with a net effect that the whole configuration has a slight bonding nature. Hence the diberyllium molecule exists (and has been observed in the gas phase). It nevertheless still has a low dissociation energy of only 59 kJ·mol^{-1}.

Dilithium

MO theory correctly predicts that dilithium is a stable molecule with bond order 1 (configuration $1\sigma_g^2 1\sigma_u^2 2\sigma_g^2$). The 1s MOs are completely filled and do not participate in bonding.

Dilithium is a gas-phase molecule with a much lower bond strength than dihydrogen because the 2s electrons are further removed from the nucleus. In a more detailed analysis which considers the environment of each orbital due to all other electrons, both the 1σ orbitals have higher energies than the 1s AO and the occupied 2σ is also higher in energy than the 2s AO.

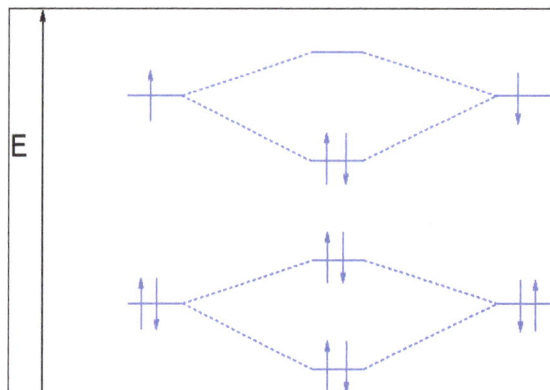

MO diagram of dilithium.

Diboron

The MO diagram for diboron (B-B, electron configuration $1\sigma_g^2 1\sigma_u^2 2\sigma_g^2 2\sigma_u^2 1\pi_u^2$) requires the introduction of an atomic orbital overlap model for p orbitals. The three dumbbell-shaped p-orbitals have equal energy and are oriented mutually perpendicularly (or orthogonally). The p-orbitals oriented in the z-direction (p_z) can overlap end-on forming a bonding (symmetrical) σ orbital and an antibonding σ^* molecular orbital. In contrast to the sigma 1s MO's, the σ 2p has some non-bonding electron density at either side of the nuclei and the σ^* 2p has some electron density between the nuclei.

The other two p-orbitals, p_y and p_x, can overlap side-on. The resulting bonding orbital has its electron density in the shape of two lobes above and below the plane of the molecule. The orbital is not symmetric around the molecular axis and is therefore a pi orbital. The antibonding pi orbital (also asymmetrical) has four lobes pointing away from the nuclei. Both p_y and p_x orbitals form a pair of pi orbitals equal in energy (degenerate) and can have higher or lower energies than that of the sigma orbital.

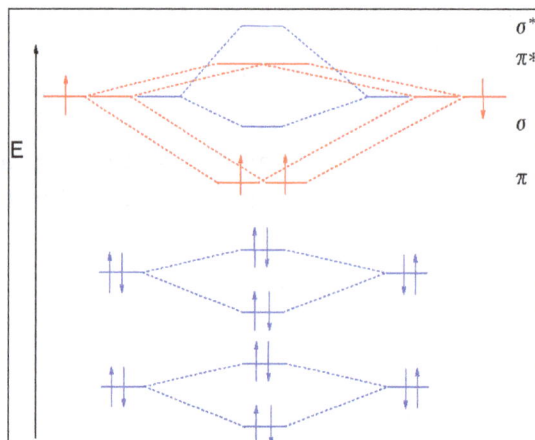

MO diagram of diboron.

In diboron the 1s and 2s electrons do not participate in bonding but the single electrons in the 2p orbitals occupy the $2\pi p_y$ and the $2\pi p_x$ MO's resulting in bond order 1. Because the electrons have equal energy (they are degenerate) diboron is a diradical and since the spins are parallel the molecule is paramagnetic.

In certain diborynes the boron atoms are excited and the bond order is 3.

Dicarbon

Like diboron, dicarbon (C-C electron configuration: $1\sigma_g^2 1\sigma_u^2 2\sigma_g^2 2\sigma_u^2 1\pi_u^4$) is a reactive gas-phase molecule. The molecule can be described as having two pi bonds but without a sigma bond.

Dinitrogen

With nitrogen, we see the two molecular orbitals mixing and the energy repulsion. This is the reasoning for the rearrangement from a more familiar diagram. Notice how the σ from the 2p behaves more non-bonding like due to mixing, same with the 2s σ. This also causes a large jump in energy in the 2p σ^* orbital. The bond order of diatomic nitrogen is three, and it is a diamagnetic molecule.

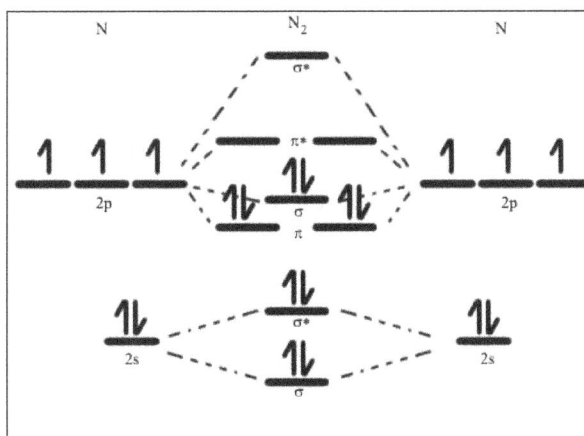

N2 Molecular Orbital Diagram.

The bond order for dinitrogen ($1\sigma_g^2 1\sigma_u^2 2\sigma_g^2 2\sigma_u^2 1\pi_u^4 3\sigma_g^2$) is three because two electrons are now also added in the 3σ MO. The MO diagram correlates with the experimental photoelectron spectrum for nitrogen. The 1σ electrons can be matched to a peak at 410 eV (broad), the $2\sigma_g$ electrons at 37 eV (broad), the $2\sigma_u$ electrons at 19 eV (doublet), the $1\pi_u^4$ electrons at 17 eV (multiplets), and finally the $3\sigma_g^2$ at 15.5 eV (sharp).

Dioxygen

Oxygen has a similar setup to H_2, but now we consider 2s and 2p orbitals. When creating the molecular orbitals from the p orbitals, notice the three atomic orbitals split into three molecular orbitals, a singly degenerate σ and a doubly degenerate π orbital. Another property we can observe by examining molecular orbital diagrams is the magnetic property of diamagnetic or paramagnetic. If all the electrons are paired, there is a slight repulsion and it is classified as diamagnetic. If unpaired electrons are present, it is attracted to a magnetic field, and therefore paramagnetic. Oxygen is an example of a paramagnetic diatomic. Also notice the bond order of diatomic oxygen is two.

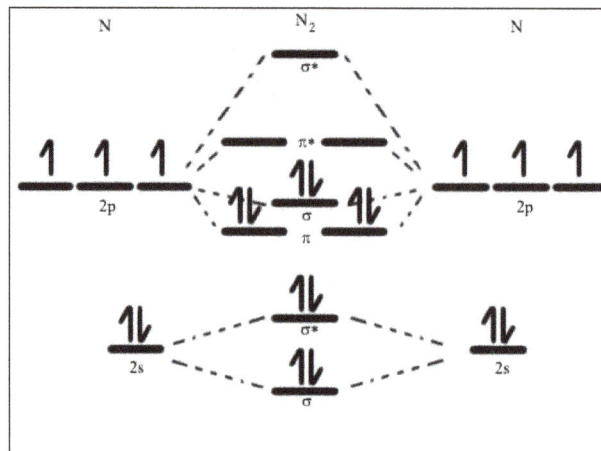

O2 Molecular Orbital Diagram.

MO treatment of dioxygen is different from that of the previous diatomic molecules because the pσ MO is now lower in energy than the 2π orbitals. This is attributed to interaction between the 2s MO and the $2p_z$ MO. Distributing 8 electrons over 6 molecular orbitals leaves the final two electrons as a degenerate pair in the 2pπ* antibonding orbitals resulting in a bond order of 2. As in diboron, these two unpaired electrons have the same spin in the ground state, which is a paramagnetic diradical triplet oxygen. The first excited state has both HOMO electrons paired in one orbital with opposite spins, and is known as singlet oxygen.

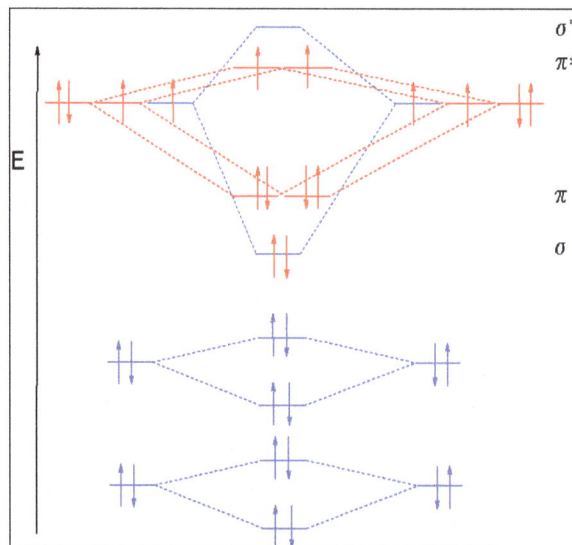

MO diagram of dioxygen triplet ground state.

The bond order decreases and the bond length increases in the order O_2^+ (112.2 pm), O_2 (121 pm), O_2^- (128 pm) and O_2^{2-} (149 pm).

Difluorine and Dineon

In difluorine two additional electrons occupy the 2pπ* with a bond order of 1. In dineon Ne_2 (as with dihelium) the number of bonding electrons equals the number of antibonding electrons and this molecule does not exist.

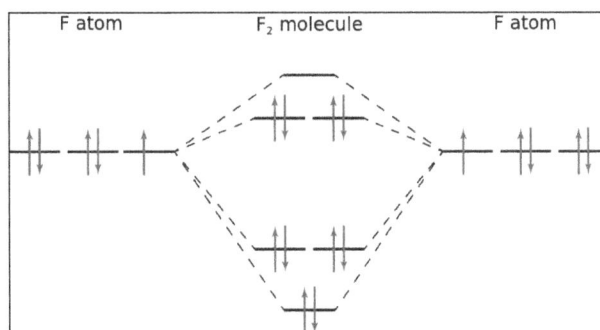

MO diagram of difluorine.

Dimolybdenum and Ditungsten

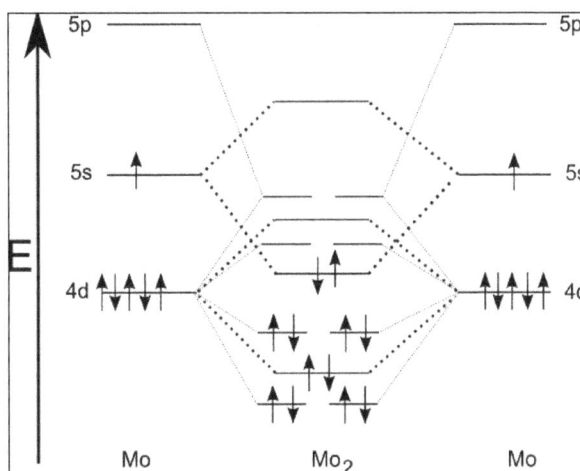

MO diagram of dimolybdenum.

Dimolybdenum (Mo_2) is notable for having a sextuple bond. This involves two sigma bonds ($4d_{z^2}$ and 5s), two pi bonds (using $4d_{xz}$ and $4d_{yz}$), and two delta bonds ($4d_{x^2-y^2}$ and $4d_{xy}$). Ditungsten (W_2) has a similar structure.

MO Energies Overview

Table gives an overview of MO energies for first row diatomic molecules calculated by the Hartree-Fock-Roothaan method, together with atomic orbital energies.

Table: Calculated MO energies for diatomic molecules in Hartrees.

	H_2	Li_2	B_2	C_2	N_2	O_2	F_2
$1\sigma_g$	-0.5969	-2.4523	-7.7040	-11.3598	-15.6820	-20.7296	-26.4289
$1\sigma_u$		-2.4520	-7.7032	-11.3575	-15.6783	-20.7286	-26.4286
$2\sigma_g$		-0.1816	-0.7057	-1.0613	-1.4736	-1.6488	-1.7620
$2\sigma_u$			-0.3637	-0.5172	-0.7780	-1.0987	-1.4997
$3\sigma_g$					-0.6350	-0.7358	-0.7504

$1\pi_u$			-0.3594	-0.4579	-0.6154	-0.7052	-0.8097
$1\pi_g$						-0.5319	-0.6682
1s (AO)	-0.5	-2.4778	-7.6953	-11.3255	-15.6289	-20.6686	-26.3829
2s (AO)		-0.1963	-0.4947	-0.7056	-0.9452	-1.2443	-1.5726
2p (AO)			-0.3099	-0.4333	-0.5677	-0.6319	-0.7300

Heteronuclear Diatomics

In heteronuclear diatomic molecules, mixing of atomic orbitals only occurs when the electronegativity values are similar. In carbon monoxide (CO, isoelectronic with dinitrogen) the oxygen 2s orbital is much lower in energy than the carbon 2s orbital and therefore the degree of mixing is low. The electron configuration $1\sigma^2 1\sigma^{*2} 2\sigma^2 2\sigma^{*2} 1\pi^4 3\sigma^2$ is identical to that of nitrogen. The g and u subscripts no longer apply because the molecule lacks a center of symmetry.

In hydrogen fluoride (HF), the hydrogen 1s orbital can mix with fluorine $2p_z$ orbital to form a sigma bond because experimentally the energy of 1s of hydrogen is comparable with 2p of fluorine. The HF electron configuration $1\sigma^2 2\sigma^2 3\sigma^2 1\pi^4$ reflects that the other electrons remain in three lone pairs and that the bond order is 1.

The more electronegative atom is the more energetically excited because it more similar in energy to its atomic orbital. This also accounts for the majority of the electron negativity residing around the more electronegative molecule. Applying the LCAO-MO method allows us to move away from a more static Lewis structure type approach and actually account for periodic trends that influence electron movement. Non-bonding orbitals refer to lone pairs seen on certain atoms in a molecule. A further understanding for the energy level refinement can be acquired by delving into quantum chemistry; the Schrödinger equation can be applied to predict movement and describe the state of the electrons in a molecule.

NO

Nitric oxide is a heteronuclear molecule that exhibits mixing. The construction of its MO diagram is the same as for the homonuclear molecules. It has a bond order of 2.5 and is a paramagnetic molecule. The energy differences of the 2s orbitals are different enough that each produces its own non-bonding σ orbitals. Notice this is a good example of making the ionized NO^+ stabilize the bond and generate a triple bond, also changing the magnetic property to diamagnetic.

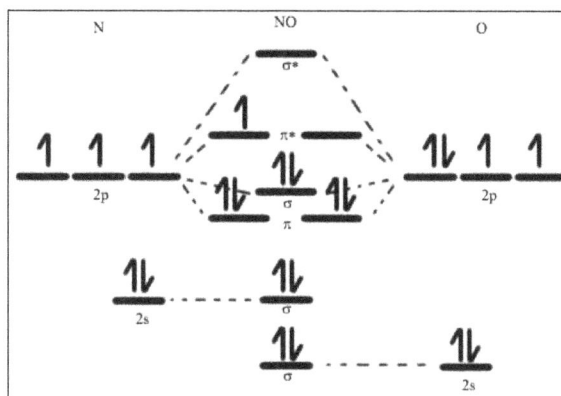

NO Molecular Orbital Diagram.

HF

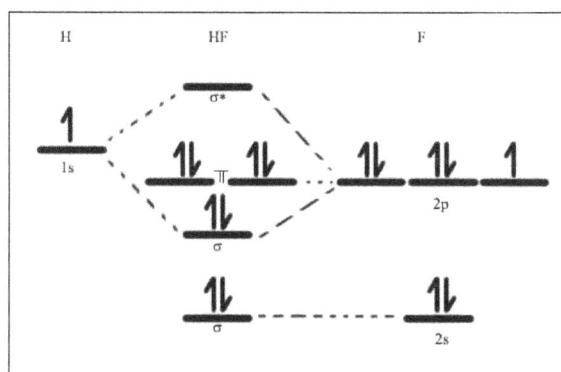

HF Molecular Orbital Diagram.

Hydrofluoric acid is another example of a homogeneous molecule. It is slightly different in that the π orbital is non-bonding, as well as the 2s σ. From the hydrogen, its valence 1s electron interacts with the 2p electrons of fluorine. This molecule is diamagnetic and has a bond order of one.

Triatomic Molecules

Carbon Dioxide

Carbon dioxide, CO_2, is a linear molecule with a total of sixteen bonding electrons in its valence shell. Carbon is the central atom of the molecule and a principal axis, the z-axis, is visualized as a single axis that goes through the center of carbon and the two oxygens atoms. For convention, blue atomic orbital lobes are positive phases, red atomic orbitals are negative phases, with respect to the wave function from the solution of the Schrödinger equation. In carbon dioxide the carbon 2s (−19.4 eV), carbon 2p (−10.7 eV), and oxygen 2p (−15.9 eV)) energies associated with the atomic orbitals are in proximity whereas the oxygen 2s energy (−32.4 eV) is different.

Carbon and each oxygen atom will have a 2s atomic orbital and a 2p atomic orbital, where the p orbital is divided into p_x, p_y, and p_z. With these derived atomic orbitals, symmetry labels are deduced with respect to rotation about the principal axis which generates a phase change, pi bond (π) or generates no phase change, known as a sigma bond (σ). Symmetry labels are further defined by whether the atomic orbital maintains its original character after an inversion about its center atom; if the atomic orbital does retain its original character it is defined gerade,*g*, or if the atomic orbital does not maintain its original character, ungerade, *u*. The final symmetry-labeled atomic orbital is now known as an irreducible representation.

Carbon dioxide's molecular orbitals are made by the linear combination of atomic orbitals of the same irreducible representation that are also similar in atomic orbital energy. Significant atomic orbital overlap explains why sp bonding may occur. Strong mixing of the oxygen 2s atomic orbital is not to be expected and are non-bonding degenerate molecular orbitals. The combination of similar atomic orbital/wave functions and the combinations of atomic orbital/wave function inverses create particular energies associated with the nonbonding (no change), bonding (lower than either parent orbital energy) and antibonding (higher energy than either parent atomic orbital energy) molecular orbitals.

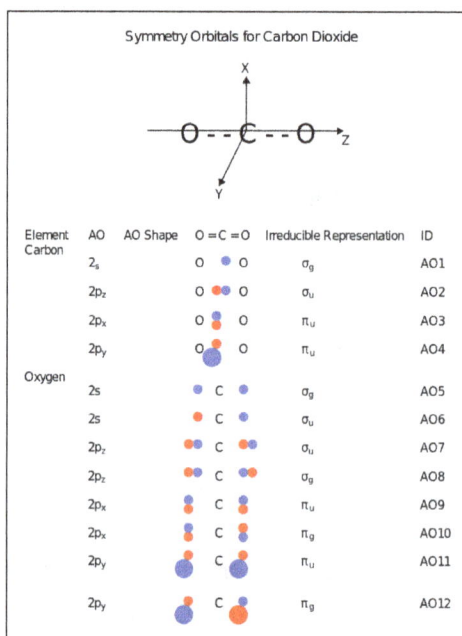

Atomic orbitals of carbon dioxide.

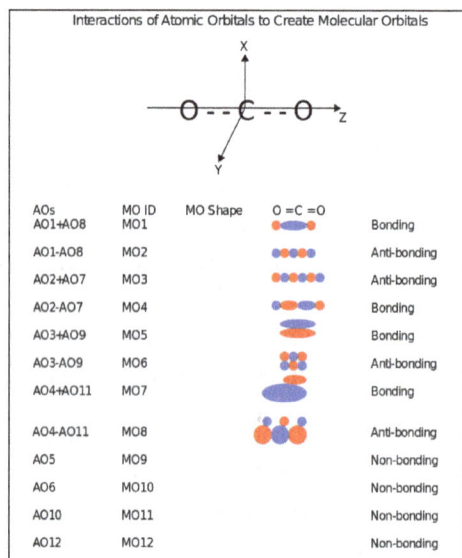

Molecular orbitals of carbon dioxide.

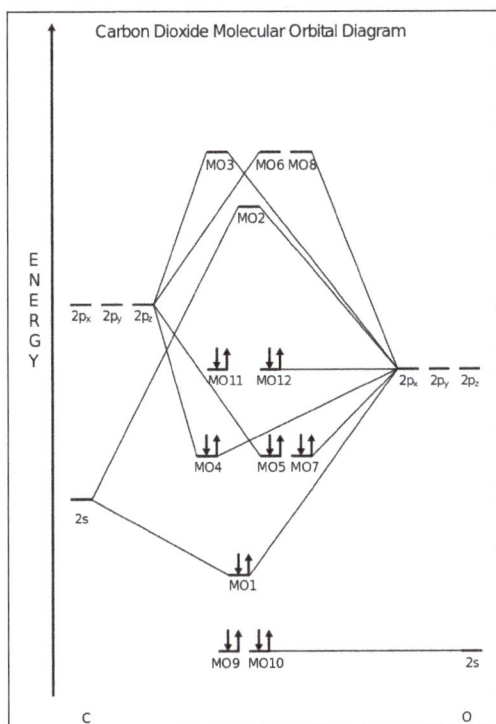

MO Diagram of carbon dioxide.

Water

For nonlinear molecules, the orbital symmetries are not σ or π but depend on the symmetry of each molecule. Water (H_2O) is a bent molecule (105°) with C_{2v} molecular symmetry. The possible orbital symmetries are listed in the table below. For example, an orbital of B_1 symmetry (called a b_1 orbital with a small b since it is a one-electron function) is multiplied by -1 under the symmetry operations C_2 (rotation about the 2-fold rotation axis) and $\sigma_v'(yz)$ (reflection in the molecular plane). It is multiplied by +1(unchanged) by the identity operation E and by $\sigma_v(xz)$ (reflection in the plane bisecting the H-O-H angle).

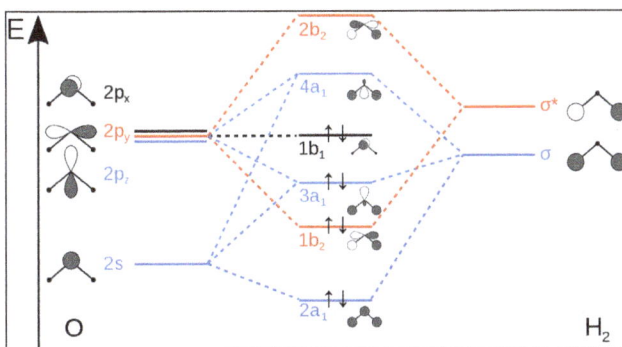

Molecular orbital diagram of water.

C2v	E	C2	σv(xz)	σv'(yz)		
A1	1	1	1	1	z	x2, y2, z2
A2	1	1	−1	−1	Rz	xy

| B1 | 1 | −1 | 1 | −1 | x, Ry | xz |
| B2 | 1 | −1 | −1 | 1 | y, Rx | yz |

The oxygen atomic orbitals are labeled according to their symmetry as a_1 for the 2s orbital and b_1 ($2p_x$), b_2 ($2p_y$) and a_1 ($2p_z$) for the three 2p orbitals. The two hydrogen 1s orbitals are premixed to form a_1 (σ) and b_2 (σ*) MO.

Mixing takes place between same-symmetry orbitals of comparable energy resulting a new set of MO's for water:

- $2a_1$ MO from mixing of the oxygen 2s AO and the hydrogen σ MO.

- $1b_2$ MO from mixing of the oxygen $2p_y$ AO and the hydrogen σ* MO.

- $3a_1$ MO from mixing of the a_1 AOs.

- $1b_1$ nonbonding MO from the oxygen $2p_x$ AO (the p-orbital perpendicular to the molecular plane).

In agreement with this description the photoelectron spectrum for water shows a sharp peak for the nonbonding $1b_1$ MO (12.6 eV) and three broad peaks for the $3a_1$ MO (14.7 eV), $1b_2$ MO (18.5 eV) and the $2a_1$ MO (32.2 eV). The $1b_1$ MO is a lone pair, while the $3a_1$, $1b_2$ and $2a_1$ MO's can be localized to give two O–H bonds and an in-plane lone pair. This MO treatment of water does not have two equivalent rabbit ear lone pairs.

Hydrogen sulfide (H_2S) too has a C_{2v} symmetry with 8 valence electrons but the bending angle is only 92°. As reflected in its photoelectron spectrum as compared to water the $5a_1$ MO (corresponding to the $3a_1$ MO in water) is stabilised (improved overlap) and the $2b_2$ MO (corresponding to the $1b_2$ MO in water) is destabilized (poorer overlap).

Electron Configuration

In quantum chemistry the electron configuration is the distribution of electrons of an atom or molecule (or other physical structure) in atomic or molecular orbitals. For example, the electron configuration of the neon atom is $1s^2\ 2s^2\ 2p^6$, using the notation explained below.

Electronic configurations describe each electron as moving independently in an orbital, in an average field created by all other orbitals. Mathematically, configurations are described by Slater determinants or configuration state functions.

According to the laws of quantum mechanics, for systems with only one electron, a level of energy is associated with each electron configuration and in certain conditions, electrons are able to move from one configuration to another by the emission or absorption of a quantum of energy, in the form of a photon.

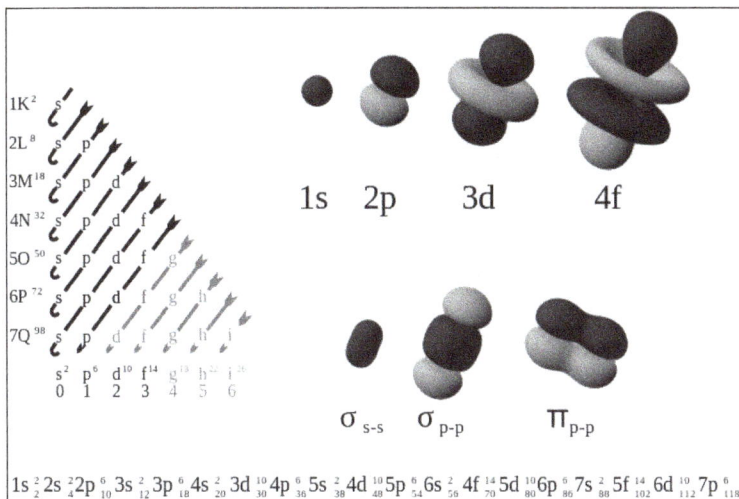

Electron atomic and molecular orbitals.

Knowledge of the electron configuration of different atoms is useful in understanding the structure of the periodic table of elements. This is also useful for describing the chemical bonds that hold atoms together. In bulk materials, this same idea helps explain the peculiar properties of lasers and semiconductors.

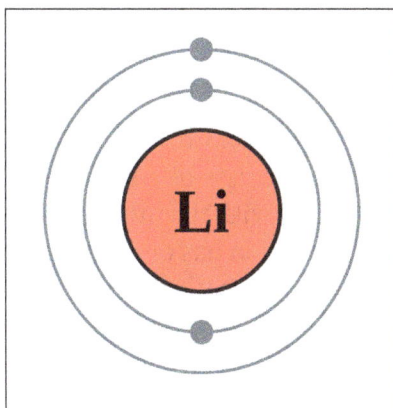

A Bohr diagram of lithium.

Shells and Subshells

	s ($\ell=0$)	p ($\ell=1$)		
	m=0	m=0	m=±1	
	s	p_z	p_x	p_y
n=1				
n=2				

Electron configuration was first conceived under the Bohr model of the atom, and it is still common to speak of shells and subshells despite the advances in understanding of the quantum-mechanical nature of electrons.

An electron shell is the set of allowed states that share the same principal quantum number, n (the number before the letter in the orbital label), that electrons may occupy. An atom's nth electron shell can accommodate $2n^2$ electrons, *e.g.* the first shell can accommodate 2 electrons, the second shell 8 electrons, the third shell 18 electrons and so on. The factor of two arises because the allowed states are doubled due to electron spin-each atomic orbital admits up to two otherwise identical electrons with opposite spin, one with a spin +1/2 (usually denoted by an up-arrow) and one with a spin −1/2 (with a down-arrow).

A subshell is the set of states defined by a common azimuthal quantum number, ℓ, within a shell. The values ℓ = 0, 1, 2, 3 correspond to the s, p, d, and f labels, respectively. For example, the 3d subshell has n = 3 and ℓ = 2. The maximum number of electrons that can be placed in a subshell is given by $2(2\ell+1)$. This gives two electrons in an s subshell, six electrons in a p subshell, ten electrons in a d subshell and fourteen electrons in an f subshell.

The numbers of electrons that can occupy each shell and each subshell arise from the equations of quantum mechanics, in particular the Pauli exclusion principle, which states that no two electrons in the same atom can have the same values of the four quantum numbers.

Notation

Physicists and chemists use a standard notation to indicate the electron configurations of atoms and molecules. For atoms, the notation consists of a sequence of atomic subshell labels (e.g. for phosphorus the sequence 1s, 2s, 2p, 3s, 3p) with the number of electrons assigned to each subshell placed as a superscript. For example, hydrogen has one electron in the s-orbital of the first shell, so its configuration is written $1s^1$. Lithium has two electrons in the 1s-subshell and one in the (higher-energy) 2s-subshell, so its configuration is written $1s^2\,2s^1$"). Phosphorus (atomic number 15) is as follows: $1s^2\,2s^2\,2p^6\,3s^2\,3p^3$.

For atoms with many electrons, this notation can become lengthy and so an abbreviated notation is used. The electron configuration can be visualized as the core electrons, equivalent to the noble gas of the preceding period, and the valence electrons: each element in a period differs only by the last few subshells. Phosphorus, for instance, is in the third period. It differs from the second-period neon, whose configuration is $1s^2\,2s^2\,2p^6$, only by the presence of a third shell. The portion of its configuration that is equivalent to neon is abbreviated as [Ne], allowing the configuration of phosphorus to be written as [Ne] $3s^2\,3p^3$ rather than writing out the details of the configuration of neon explicitly. This convention is useful as it is the electrons in the outermost shell that most determine the chemistry of the element.

For a given configuration, the order of writing the orbitals is not completely fixed since only the orbital occupancies have physical significance. For example, the electron configuration of the titanium ground state can be written as either [Ar] $4s^2\,3d^2$ or [Ar] $3d^2\,4s^2$. The first notation follows the order based on the Madelung rule for the configurations of neutral atoms; 4s is filled before 3d in the sequence Ar, K, Ca, Sc, Ti. The second notation groups all orbitals with the same value of n together, corresponding to the "spectroscopic" order of orbital energies that is the reverse of the order in which electrons are removed from a given atom to form positive ions; 3d is filled before 4s in the sequence Ti^{4+}, Ti^{3+}, Ti^{2+}, Ti^{+}, Ti.

The superscript 1 for a singly occupied subshell is not compulsory; for example aluminium may be written as either [Ne] $3s^2$ $3p^1$ or [Ne] $3s^2$ 3p. It is quite common to see the letters of the orbital labels (s, p, d, f) written in an italic or slanting typeface, although the International Union of Pure and Applied Chemistry (IUPAC) recommends a normal typeface. The choice of letters originates from a now-obsolete system of categorizing spectral lines as "sharp", "principal", "diffuse" and "fundamental" (or "fine"), based on their observed fine structure: their modern usage indicates orbitals with an azimuthal quantum number, l, of 0, 1, 2 or 3 respectively. After "f", the sequence continues alphabetically "g", "h", "i"... ($l = 4, 5, 6..$)., skipping "j", although orbitals of these types are rarely required.

The electron configurations of molecules are written in a similar way, except that molecular orbital labels are used instead of atomic orbital labels.

Energy-ground State and Excited States

The energy associated to an electron is that of its orbital. The energy of a configuration is often approximated as the sum of the energy of each electron, neglecting the electron-electron interactions. The configuration that corresponds to the lowest electronic energy is called the ground state. Any other configuration is an excited state.

As an example, the ground state configuration of the sodium atom is $1s^2 2s^2 2p^6 3s^1$, as deduced from the Aufbau principle. The first excited state is obtained by promoting a 3s electron to the 3p orbital, to obtain the $1s^2 2s^2 2p^6 3p^1$ configuration, abbreviated as the 3p level. Atoms can move from one configuration to another by absorbing or emitting energy. In a sodium-vapor lamp for example, sodium atoms are excited to the 3p level by an electrical discharge, and return to the ground state by emitting yellow light of wavelength 589 nm.

Usually, the excitation of valence electrons (such as 3s for sodium) involves energies corresponding to photons of visible or ultraviolet light. The excitation of core electrons is possible, but requires much higher energies, generally corresponding to x-ray photons. This would be the case for example to excite a 2p electron of sodium to the 3s level and form the excited $1s^2 2s^2 2p^5 3s^2$ configuration.

Atoms: Aufbau Principle and Madelung Rule

The Aufbau principle (from the German *Aufbau*, "building up, construction") was an important part of Bohr's original concept of electron configuration. It may be stated as:

A maximum of two electrons are put into orbitals in the order of increasing orbital energy: the lowest-energy orbitals are filled before electrons are placed in higher-energy orbitals.

The principle works very well (for the ground states of the atoms) for the first 18 elements, then decreasingly well for the following 100 elements. The modern form of the Aufbau principle describes an order of orbital energies given by Madelung's rule (or Klechkowski's rule). This rule was first stated by Charles Janet in 1929, rediscovered by Erwin Madelung in 1936, and later given a theoretical justification by V.M. Klechkowski:

- Orbitals are filled in the order of increasing $n+l$;

- Where two orbitals have the same value of $n+l$, they are filled in order of increasing n.

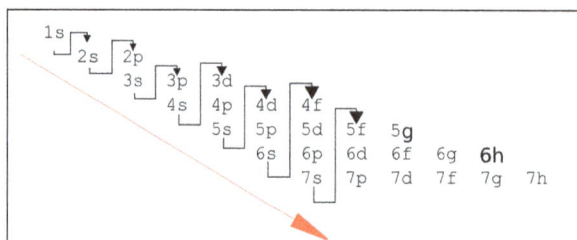

The approximate order of filling of atomic orbitals, following the arrows from 1s to 7p.
(After 7p the order includes orbitals outside the range of the diagram, starting with 8s).

This gives the following order for filling the orbitals:

1s, 2s, 2p, 3s, 3p, 4s, 3d, 4p, 5s, 4d, 5p, 6s, 4f, 5d, 6p, 7s, 5f, 6d, 7p, (8s, 5g, 6f, 7d, 8p, and 9s).

In this list the orbitals in parentheses are not occupied in the ground state of the heaviest atom now known (Og, Z = 118).

The Aufbau principle can be applied, in a modified form, to the protons and neutrons in the atomic nucleus, as in the shell model of nuclear physics and nuclear chemistry.

Periodic Table

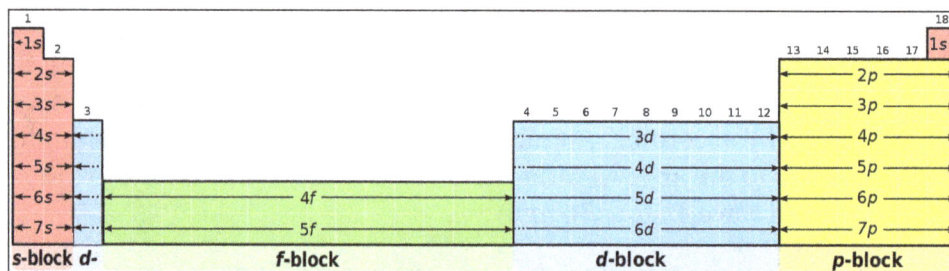

Electron configuration table.

The form of the periodic table is closely related to the electron configuration of the atoms of the elements. For example, all the elements of group 2 have an electron configuration of [E] ns² (where [E] is an inert gas configuration), and have notable similarities in their chemical properties. In general, the periodicity of the periodic table in terms of periodic table blocks is clearly due to the number of electrons (2, 6, 10, 14..). needed to fill s, p, d, and f subshells.

The outermost electron shell is often referred to as the "valence shell" and (to a first approximation) determines the chemical properties. It should be remembered that the similarities in the chemical properties were remarked on more than a century before the idea of electron configuration. It is not clear how far Madelung's rule explains (rather than simply describes) the periodic table, although some properties (such as the common +2 oxidation state in the first row of the transition metals) would obviously be different with a different order of orbital filling.

Shortcomings of the Aufbau Principle

The Aufbau principle rests on a fundamental postulate that the order of orbital energies is fixed, both for a given element and between different elements; in both cases this is only approximately true. It considers atomic orbitals as "boxes" of fixed energy into which can be placed two electrons and no more. However, the energy of an electron "in" an atomic orbital depends on the energies of all the other electrons of the atom (or ion, or molecule, etc). There are no "one-electron solutions" for systems of more than one electron, only a set of many-electron solutions that cannot be calculated exactly (although there are mathematical approximations available, such as the Hartree–Fock method).

The fact that the Aufbau principle is based on an approximation can be seen from the fact that there is an almost-fixed filling order at all, that, within a given shell, the s-orbital is always filled before the p-orbitals. In a hydrogen-like atom, which only has one electron, the s-orbital and the p-orbitals of the same shell have exactly the same energy, to a very good approximation in the absence of external electromagnetic fields. (However, in a real hydrogen atom, the energy levels are slightly split by the magnetic field of the nucleus, and by the quantum electrodynamic effects of the Lamb shift).

Ionization of the Transition Metals

The naïve application of the Aufbau principle leads to a well-known paradox (or apparent paradox) in the basic chemistry of the transition metals. Potassium and calcium appear in the periodic table before the transition metals, and have electron configurations [Ar] $4s^1$ and [Ar] $4s^2$ respectively, i.e. the 4s-orbital is filled before the 3d-orbital. This is in line with Madelung's rule, as the 4s-orbital has $n+l = 4$ ($n = 4$, $l = 0$) while the 3d-orbital has $n+l = 5$ ($n = 3$, $l = 2$). After calcium, most neutral atoms in the first series of transition metals (Sc-Zn) have configurations with two 4s electrons, but there are two exceptions. Chromium and copper have electron configurations [Ar] $3d^5$ $4s^1$ and [Ar] $3d^{10}$ $4s^1$ respectively, i.e. one electron has passed from the 4s-orbital to a 3d-orbital to generate a half-filled or filled subshell. In this case, the usual explanation is that "half-filled or completely filled subshells are particularly stable arrangements of electrons".

The apparent paradox arises when electrons are *removed* from the transition metal atoms to form ions. The first electrons to be ionized come not from the 3d-orbital, as one would expect if it were "higher in energy", but from the 4s-orbital. This interchange of electrons between 4s and 3d is found for all atoms of the first series of transition metals. The configurations of the neutral atoms (K, Ca, Sc, Ti, V, Cr, ..). usually follow the order 1s, 2s, 2p, 3s, 3p, 4s, 3d, ...; however the successive stages of ionization of a given atom (such as Fe^{4+}, Fe^{3+}, Fe^{2+}, Fe^+, Fe) usually follow the order 1s, 2s, 2p, 3s, 3p, 3d, 4s, ...

This phenomenon is only paradoxical if it is assumed that the energy order of atomic orbitals is fixed and unaffected by the nuclear charge or by the presence of electrons in other orbitals. If that were the case, the 3d-orbital would have the same energy as the 3p-orbital, as it does in hydrogen, yet it clearly doesn't. There is no special reason why the Fe^{2+} ion should have the same electron configuration as the chromium atom, given that iron has two more protons in its nucleus than chromium, and that the chemistry of the two species is very different. Melrose and Eric Scerri have analyzed the changes of orbital energy with orbital occupations in terms of the two-electron repulsion integrals of the Hartree-Fock method of atomic structure calculation.

Similar ion-like $3d^x4s^0$ configurations occur in transition metal complexes as described by the simple crystal field theory, even if the metal has oxidation state 0. For example, chromium hexacarbonyl can be described as a chromium atom (not ion) surrounded by six carbon monoxide ligands. The electron configuration of the central chromium atom is described as $3d^6$ with the six electrons filling the three lower-energy d orbitals between the ligands. The other two d orbitals are at higher energy due to the crystal field of the ligands. This picture is consistent with the experimental fact that the complex is diamagnetic, meaning that it has no unpaired electrons. However, in a more accurate description using molecular orbital theory, the d-like orbitals occupied by the six electrons are no longer identical with the d orbitals of the free atom.

Other Exceptions to Madelung's Rule

There are several more exceptions to Madelung's rule among the heavier elements, and as atomic number increases it becomes more and more difficult to find simple explanations such as the stability of half-filled subshells. It is possible to predict most of the exceptions by Hartree–Fock calculations, which are an approximate method for taking account of the effect of the other electrons on orbital energies. For the heavier elements, it is also necessary to take account of the effects of special relativity on the energies of the atomic orbitals, as the inner-shell electrons are moving at speeds approaching the speed of light. In general, these relativistic effects tend to decrease the energy of the s-orbitals in relation to the other atomic orbitals. The table below shows the ground state configuration in terms of orbital occupancy, but it does not show the ground state in terms of the sequence of orbital energies as determined spectroscopically. For example, in the transition metals, the 4s orbital is of a higher energy than the 3d orbitals; and in the lanthanides, the 6s is higher than the 4f and 5d. The ground states can be seen in the Electron configurations of the elements.

The electron-shell configuration of elements beyond hassium has not yet been empirically verified, but they are expected to follow Madelung's rule without exceptions until element 120. Beyond element 120, Madelung's rule is expected to stop holding altogether due to the closeness in energy of the 5g, 6f, 7d, and $8p_{1/2}$ orbitals.

Electron Configuration in Molecules

In molecules, the situation becomes more complex, as each molecule has a different orbital structure. The molecular orbitals are labelled according to their symmetry, rather than the atomic orbital labels used for atoms and monatomic ions: hence, the electron configuration of the dioxygen molecule, O_2, is written $1\sigma_g^2 1\sigma_u^2 2\sigma_g^2 2\sigma_u^2 3\sigma_g^2 1\pi_u^4 1\pi_g^2$, or equivalently $1\sigma_g^2 1\sigma_u^2 2\sigma_g^2 2\sigma_u^2 1\pi_u^4 3\sigma_g^2 1\pi_g^2$. The term $1\pi_g^2$ represents the two electrons in the two degenerate π^*-orbitals (antibonding). From Hund's rules, these electrons have parallel spins in the ground state, and so dioxygen has a net magnetic moment (it is paramagnetic). The explanation of the paramagnetism of dioxygen was a major success for molecular orbital theory.

The electronic configuration of polyatomic molecules can change without absorption or emission of a photon through vibronic couplings.

Electron Configuration in Solids

In a solid, the electron states become very numerous. They cease to be discrete, and effectively blend into continuous ranges of possible states (an electron band). The notion of electron configuration ceases to be relevant, and yields to band theory.

Applications

The most widespread application of electron configurations is in the rationalization of chemical properties, in both inorganic and organic chemistry. In effect, electron configurations, along with some simplified form of molecular orbital theory, have become the modern equivalent of the valence concept, describing the number and type of chemical bonds that an atom can be expected to form.

This approach is taken further in computational chemistry, which typically attempts to make quantitative estimates of chemical properties. For many years, most such calculations relied upon the "linear combination of atomic orbitals" (LCAO) approximation, using an ever-larger and more complex basis set of atomic orbitals as the starting point. The last step in such a calculation is the assignment of electrons among the molecular orbitals according to the Aufbau principle. Not all methods in calculational chemistry rely on electron configuration: density functional theory (DFT) is an important example of a method that discards the model.

For atoms or molecules with more than one electron, the motion of electrons are correlated and such a picture is no longer exact. A very large number of electronic configurations are needed to exactly describe any multi-electron system, and no energy can be associated with one single configuration. However, the electronic wave function is usually dominated by a very small number of configurations and therefore the notion of electronic configuration remains essential for multi-electron systems.

A fundamental application of electron configurations is in the interpretation of atomic spectra. In this case, it is necessary to supplement the electron configuration with one or more term symbols, which describe the different energy levels available to an atom. Term symbols can be calculated for any electron configuration, not just the ground-state configuration listed in tables, although not all the energy levels are observed in practice. It is through the analysis of atomic spectra that the ground-state electron configurations of the elements were experimentally determined.

The Hückel Approximation

Consider a conjugated molecule i.e. a molecule with alternating double and single bonds, as shown in figure.

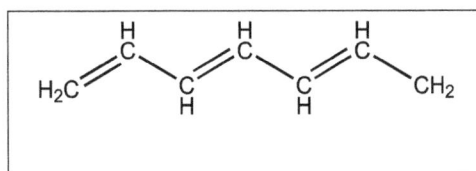

The Hückel approximation is used to determine the energies and shapes of the π molecular orbitals. In other words, the Hückel approximation assumes that the electrons in the π bonds "feel" an electrostatic potential due to the entire σ (sigma) bonding framework in the molecule (i.e. it focuses only on the formation of π bonds, given that the σ bonding framework has already been formed). Let's look at the simplest example, ethylene $CH_2=CH_2$.

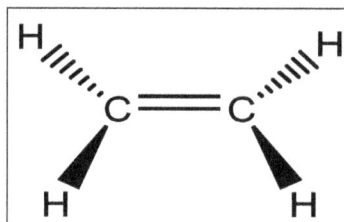

In ethylene the carbon atoms are sp^2 hybridized and the sigma bonding framework consists of the sp^2-sp^2 overlap between the two carbon atoms and four sp^2-s overlaps between the sp^2 orbitals of the carbon atoms and the s orbitals of the hydrogen atoms. On each carbon atom is a 2py orbital lying perpendicular to the plane containing all the σ bonds (C-C and C-H σ bonds, see figure). Each of the 2py orbital contains a single electron.

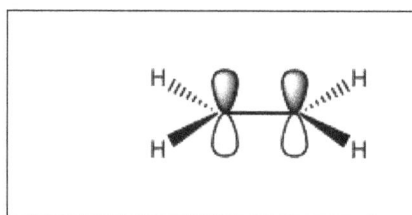

(The orientation of two p orbitals overlapping to form a σ bond is perpendicular to that shown in figure. In other words, for two p orbitals to overlap and form a σ bond, the two p orbitals must lie along the inter-nuclear axis connecting the nuclei of the two atoms, as shown in figure. The resulting σ bond is formed along the bond axis).

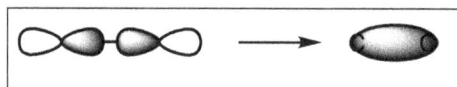

The two $2p_y$ atomic orbitals on the carbon atoms can overlap to form π molecular orbitals. As with overlap of atomic orbitals to form σ molecular orbitals, when the two $2p_y$ atomic orbitals overlap two π molecular orbitals are formed. The Hückel approximation is used to determine the energies and shapes of these two π orbitals. The energy level diagram of the π molecular orbitals, as determined by the Hückel approximation, is shown in figure. In this figure, the dotted line denotes the energy of the $2p_y$ atomic orbitals. As shown in figure, one of the π molecular orbitals is lower in energy relative to the $2p_y$ orbitals and the other is higher in energy relative to the $2p_y$ orbitals.

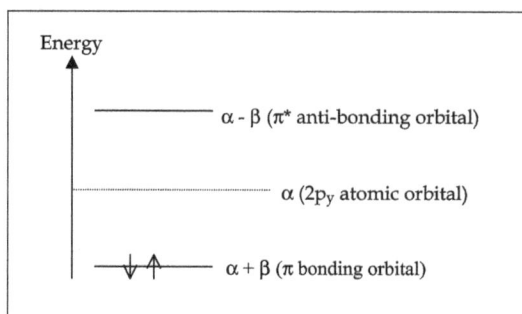

The π orbital lower in energy is called the π bonding orbital (analogous to the σ bonding orbital) and the higher energy π orbital is the π* anti-bonding orbital (analogous to the σ* antibonding orbital). If the energy of the 2py orbitals is α, the Hückel approximation predicts that the π bonding orbital is lower in energy by a value called β and π^* is higher in energy by $-\beta$. Since β turns out to be negative (β = -75 kJ/mol), the energy of the π orbital is $\alpha + \beta$ and that of π^* is $\alpha - \beta$.

Since each $2p_y$ orbital is occupied by one electron, on forming the π bonds, both electrons now occupy the π bonding molecular orbital (with opposite spins). Hence, the energy of each electron occupying the π bonding orbital is lowered by a value of β relative to the $2p_y$ orbitals. So on forming the π bonds the total energy of the ethylene molecule is lowered by 2β ($\beta + \beta$, for each π electron) relative to the energy of the ethylene molecule with two 2py atomic orbitals. Hence, forming the π bonds makes the ethylene molecule more stable.

We can pictorially represent these two molecular orbitals as follows:

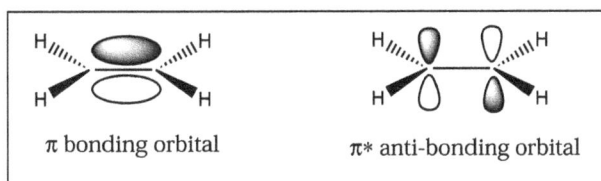

π bonding orbital π^* anti-bonding orbital

From the figure of the π bonding orbital, you can immediately see that the effect of forming the π bonding molecular orbital is that the electrons, originally in each of the $2p_y$ atomic orbitals located over single carbon atoms, are now spread out or "delocalized" in a molecular orbital over two carbon atoms. This delocalization of the position of the electrons results in lowering the energy of the π bonding orbital relative to the $2p_y$ atomic orbitals.

Let's now look at the π molecular orbitals of a slightly larger molecule 1,3-butadiene.

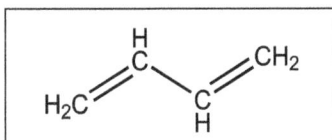

Similar to ethylene, each of the carbon atoms has a $2p_y$ atomic orbital, lying perpendicular to the σ bond framework (all C-C and C-H σ bonds), with an electron in each of these atomic orbitals. These four $2p_y$ orbitals can overlap to form π bonds. Once again, using the Hückel approximation we can determine the energies and shapes of these π molecular orbitals. As we can see in figure, combining these four $2p_y$ atomic orbitals, we end up with four π molecular orbitals.

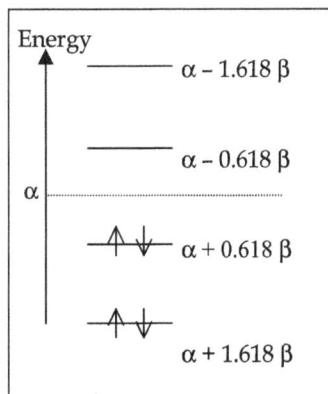

Once again, we will reference the energies of these orbitals with respect to α, the energy of the 2p$_y$ atomic orbitals. We see that two π molecular orbitals lie below α (these are the π bonding orbitals located -1.618β and -0.618β below α) and two lie above α (these are the π^* anti-bonding orbitals located 1.618β and 0.618β above α). The four electrons from each of the 2py orbitals occupy the lowest two π bonding orbitals, in pairs and with opposite spin.

Since the four π electrons occupy molecular orbitals at energies lower than the 2p$_y$ atomic orbitals, we expect that formation of the π bonds lowers the energy of the butadiene molecule. In fact, it lowers the energy by 2(1.618β) + 2(0.618β) = 4.472β (each electron occupying the lowest π orbital has an energy 1.618β lower than α and each electron in the next π orbital has an energy 0.618β lower than α). So once again forming π bonds lowers the energy of the molecule. As with ethylene, the energy of the molecule is lowered since the π electrons are now delocalized over the entire butadiene molecule instead of being localized in a 2py atomic orbital. Figure shows the shape of just the lowest π bonding molecular orbital with energy α + 1.618 β.

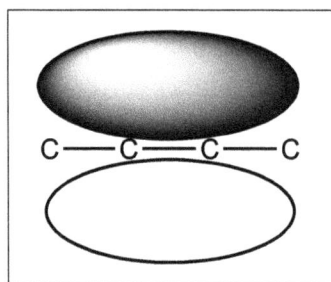

Effect of the Length of the Conjugated Molecule on Delocalization.

As we saw above, the butadiene molecule is more stable by 4.472β with respect to the energies of four 2py atomic orbital. Let's compare this energy with the sum of the energies of two isolated ethylene molecules (once again relative to the energies of four 2py atomic orbitals). Each ethylene molecule is stabilized by 2β (relative to the energies of four 2py atomic orbitals). So two isolated ethylene molecules are stabilized by a value of 4β (relative to the energies of four 2py atomic orbitals). Comparing the stabilization energy of two ethylene molecules due to π bonding (4β) to that of one butadiene molecule (4.472 β), we find that the butadiene molecule is more stable by 0.472β. Why is this so? This extra stabilization in the butadiene molecule is once again due to delocalization of the π electrons over the length of the molecule. In general, the longer the length of the conjugated molecule, the larger is the degree of stabilization due to delocalization.

Hückel Molecular Orbital Theory

In general, the vast majority polyatomic molecules can be thought of as consisting of a collection of twoelectron bonds between pairs of atoms. So the qualitative picture of s and pbonding and antibonding orbitals that we developed for a diatomic like CO can be carried over give a qualitative starting point for describing the C=O bond in acetone, for example. One place where this qualitative picture is extremely useful is in dealing with conjugated systems   that is, molecules that contain a series of alternating double/single bonds in their Lewis structure like 1,3,5hexatriene.

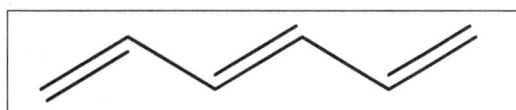

Now, you may have been taught in previous courses that because there are other resonance structures you can draw for this molecule, such as:

That it is better to think of the molecule as having a series of bonds of order 1 ½ rather than 2/1/2/1/... MO theory actually predicts this behavior, and this prediction is one of the great successes of MO theory as a descriptor of chemistry.

Conjugated molecules of tend to be planar, so that we can place all the atoms in the xy plane. Thus, the molecule will have reflection symmetry about the zaxis.

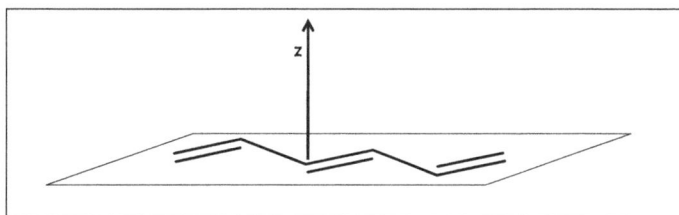

Now, for diatomics, we had reflection symmetry about x and y and this gave rise to p_x and p_y orbitals that were odd with respect to reflection and σ orbitals that were even. In the same way, for planar conjugated systems the orbitals will separate into σ orbitals that are even with respect to reflection and p_z orbitals that are odd with respect to reflection about z. These P_z orbitals will be linear combinations of the p_z orbitals on each carbon atom:

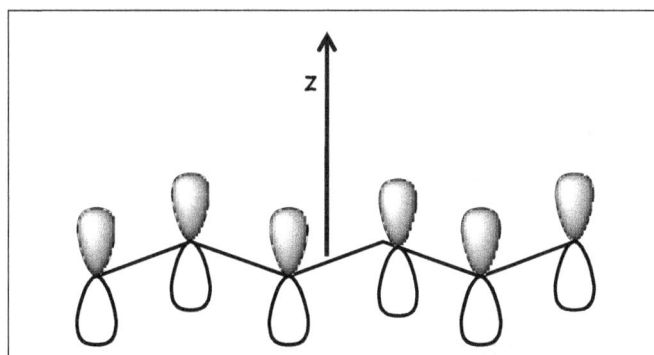

In trying to understand the chemistry of these compounds, it makes sense to focus our attention on these p_z orbitals and ignore the σ orbitals. The p_z orbitals turn out to be the highest occupied orbitals, with the σ orbitals being more strongly bound. Thus, the forming and breaking of bonds – as implied by our resonance structures – will be easier if we talk about making and breaking π bonds rather than σ. Thus, at a basic level, we can ignore the existence of the sorbitals and deal only with the porbitals in a qualitative MO theory of conjugated systems. This is the basic approximation of Hückel theory, which can be outlined in the standard 5 steps of MO theory.

- Define a basis of atomic orbitals. Here, since we are only interested in the pz orbitals, we will be able to write out MOs as linear combinations of the p_z orbitals. If we assume there are N carbon atoms, each contributes a p_z orbital and we can write the mth MOs as:

$$\pi^\mu = \sum_{i=1}^N c_i^\mu P_z^i$$

- Compute the relevant matrix representations. Hückel makes some radical approximations at this step that make the algebra much simpler without changing the qualitative answer. We have to compute two matrices, H and S which will involve integrals between p_z orbitals on different carbon atoms.

$$H_{ij} = \int p_z^i \hat{H} p_z^j d\tau \quad S_{ij} = \int p_z^i p_z^j d\tau$$

The first approximation we make is that the pz orbitals are orthonormal. This means that:

$$S_{ij} = \begin{cases} 1 & i=j \\ 0 & i=j \end{cases}$$

Equivalently, this means S is the identity matrix, which reduces our generalized eigenvalue problem to a normal eigenvalue problem:

$$\mathrm{H} \cdot c^\alpha = E_\alpha \mathrm{S} \cdot c^\mu \implies \mathrm{H} \cdot \mathrm{c}^\mu = E_\mu c^\mu$$

The second approximation we make is to assume that any Hamiltonian integrals vanish if they involve atoms i,j that are not nearest neighbors. This makes some sense, because when the p_z orbitals are far apart they will have very little spatial overlap, leading to an integrand that is nearly zero everywhere. We note also that the diagonal (i=j) terms must all be the same because they involve the average energy of an electron in a carbon p_z orbital:

$$H_{ij} = \int p_z^i \hat{H} p_z^I d\tau \equiv \alpha$$

Because it describes the energy of an electron on a single carbon, a is often called the onsite energy. Meanwhile, for any two nearest neighbors, the matrix element will also be assumed to be constant:

$$H_{ij} = \int p_z^i \hat{H} p_z^j d\tau \equiv \beta \quad i, j \text{ neigbors}$$

This last approximation is good as long as the CC bond lengths in the molecule are all nearly equal. If there is significant bond length alternation (e.g. single/double/single) then this approximation can be relaxed to allow b to depend on the CC bond distance. As we will see, b allows us to describe the electron delocalization that comes from multiple resonance structures and hence it is often called a resonance integral. There is some debate about what the "right" values for the a, b parameters are, but one good choice is $\alpha = 11.2$ eV and $\beta = .7$ eV.

- Solve the generalized eigenvalue problem. Here, we almost always need to use a computer. But because the matrices are so simple, we can usually find the eigenvalues and eigenvectors very quickly.

- Occupy the orbitals according to a stick diagram. At this stage, we note that from our $N\,p_z$ orbitals we will obtain N p orbitals. Further, each carbon atom has one free valence electron to contribute, for a total of N electrons that will need to be accounted for (assuming the molecule is neutral). Accounting for spin, then, there will be $N/2$ occupied molecular orbitals and $N/2$ unoccupied ones. For the ground state, we of course occupy the lowest energy orbitals.

- Compute the energy. Being a very approximate form of MO theory, Hückel uses the non-interacting electron energy expression:

$$E_{tot} = \sum_{i=1}^{N} E_i$$

Where E_i are the MO eigenvalues determined in the third step.

To illustrate how we apply Hückel in practice, let's work out the energy of benzene as an example:

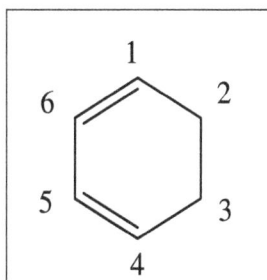

- Each of the MOs is a linear combination of 6 pz orbitals:

$$\psi^\mu = \sum_{i=1}^{6} c_i^\mu P_z^i \rightarrow c^\mu = \begin{pmatrix} c_1^\mu \\ c_2^\mu \\ c_3^\mu \\ c_4^\mu \\ c_5^\mu \\ c_6^\mu \end{pmatrix}$$

- It is relatively easy to work out the Hamiltonian. It is a 6by6 matrix. The first rule implies that every diagonal element is α:

$$H = \begin{pmatrix} \alpha & & & & & \\ & \alpha & & & & \\ & & \alpha & & & \\ & & & \alpha & & \\ & & & & \alpha & \\ & & & & & \alpha \end{pmatrix}$$

The only other nonzero terms will be between neighbors: 12, 23, 34, 45, 56 and 61. All these elements are equal to β:

$$H = \begin{pmatrix} \alpha & \beta & & & & \beta \\ \beta & \alpha & \beta & & & \\ & \beta & \alpha & \beta & & \\ & & \beta & \alpha & \beta & \\ & & & \beta & \alpha & \beta \\ \beta & & & & \beta & \alpha \end{pmatrix}$$

All the rest of the elements involve nonnearest neighbors and so are zero:

$$H = \begin{pmatrix} \alpha & \beta & 0 & 0 & 0 & \beta \\ \beta & \alpha & \beta & 0 & 0 & 0 \\ 0 & \beta & \alpha & \beta & 0 & 0 \\ 0 & 0 & \beta & \alpha & \beta & 0 \\ 0 & 0 & 0 & \beta & \alpha & \beta \\ \beta & 0 & 0 & 0 & \beta & \alpha \end{pmatrix}$$

• Finding the eigenvalues of H is easy with a computer. We find 4 distinct energies:

$$E_6 = \alpha - 2\beta$$
$$E_4 = E_5 = \alpha - \beta$$
$$E_2 = E_3 = \alpha + \beta$$
$$E_1 = \alpha + 2\beta$$

The lowest and highest energies are nondegenerate. The second/third and fourth/fifth energies are degenerate with one another. With a little more work we can get the eigenvectors. They are:

$$C^6 = \frac{1}{\sqrt{6}}\begin{pmatrix} +1 \\ -1 \\ +1 \\ -1 \\ +1 \\ -1 \end{pmatrix} \quad C^5 = \frac{1}{\sqrt{12}}\begin{pmatrix} +1 \\ -2 \\ +1 \\ +1 \\ -2 \\ +1 \end{pmatrix} \quad C^4 = \frac{1}{\sqrt{4}}\begin{pmatrix} +1 \\ 0 \\ -1 \\ +1 \\ 0 \\ -1 \end{pmatrix} \quad C^3 = \frac{1}{\sqrt{4}}\begin{pmatrix} +1 \\ 0 \\ -1 \\ -1 \\ 0 \\ +1 \end{pmatrix} \quad C^2 = \frac{1}{\sqrt{12}}\begin{pmatrix} +1 \\ +2 \\ +1 \\ -1 \\ -2 \\ -1 \end{pmatrix} \quad C^1 = \frac{1}{\sqrt{6}}\begin{pmatrix} +1 \\ +1 \\ +1 \\ +1 \\ +1 \\ +1 \end{pmatrix}$$

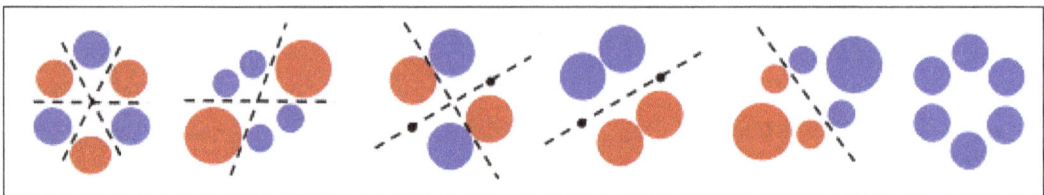

The pictures at the bottom illustrate the MOs by denting positive (negative) lobes by circles whose size corresponds to the weight of that particular p_z orbital in the MO. The resulting phase pattern is very reminiscent of a particle on a ring, where we saw that the ground state had no nodes, the first and second excited states were degenerate (sine and cosine) and had one node, the third and fourth were degenerate with two nodes. The one difference is that, in benzene the fifth excited state is the only one with three nodes, and it is nondegenerate.

- There are 6 p electrons in benzene, so we doubly occupy the first 3 MOs:

$$E_6 = \alpha - 2\beta$$
$$E_4 = E_5 = \alpha - \beta$$
$$E_2 = E_3 = \alpha + \beta$$
$$E_1 = \alpha + 2\beta$$

- The Hückel energy of benzene is then:

$$E = 2E_1 + 2E_2 + 2E_3 = 6\alpha + 8\beta$$

Now, we get to the interesting part. What does this tell us about the bonding in benzene? Well, first we note that benzene is somewhat more stable than a typical system with three double bonds would be. If we do Hückel theory for ethylene, we find that a single ethylene double bond has an energy.

$$E_{C=C} = 2\alpha + 2\beta$$

Thus, if benzene simply had three double bonds, we would expect it to have a total energy of:

$$E = 3E_{C=C} = 6\alpha + 6\beta,$$

which is off by 2β. We recall that b is negative, so that the **p**electrons in benzene are more stable than a collection of three double bonds. We call this *aromatic stabilization*, and Hückel theory predicts a similar stabilization of other cyclic conjugated systems with 4N+2 electrons. This energetic stabilization explains in part why benzene is so unreactive as compared to other unsaturated hydrocarbons.

We can go one step further in our analysis and look at the bond order. In Hückel theory the bond order can be defined as:

$$O_{ij} = \sum_{\mu=1}^{occ} c_i^\mu \, c_j^\mu$$

This definition incorporates the idea that, if molecular orbital m has a bond between the ith and j[th] carbons, then the coefficients of the MO on those carbons should both have the same sign (e.g. we have $p_z^i - p_z^{j\,j}$). If the orbital is antibonding between i and j, the coefficients should have opposite signs(e.g. we have $p_z^i \quad p_z^j$). The summand above reflects this because.

$$c_i^{\mu} c_j^{\mu} > 0 \qquad \text{if } c_i^{\mu}, c_j^{\mu} \quad \text{have same sign}$$

$$c_i^{\mu} c_j^{\mu} > 0 \qquad \text{if } c_i^{\mu}, c_j^{\mu} \quad \text{have opposite sign}$$

Thus the formula gives a positive contribution for bonding orbitals and a negative contribution for antibonding. The summation over the occupied orbitals just sums up the bonding or antibonding contributions from all the occupied MOs for the particular ijpair of carbons to get the total bond order. Note that, in this summation, a doubly occupied orbital will appear twice. Applying this formula to the 12 bond in benzene, we find that:

$$O_{12} \equiv 2c_1^{\mu=1} c_2^{\mu=1} + 2c_1^{\mu=2} c_2^{\mu=2} + 2c_1^{\mu=3} c_2^{\mu=3}$$

$$= 2\left(\frac{+1}{\sqrt{6}}\right) \times \left(\frac{+1}{\sqrt{6}}\right) + 2\left(\frac{+1}{\sqrt{12}}\right) \times \left(\frac{+2}{\sqrt{12}}\right) + 2\left(\frac{+1}{\sqrt{4}}\right) \times \left(\frac{0}{\sqrt{4}}\right)$$

$$= 2\frac{1}{6} + 2\frac{2}{12} = \frac{2}{3}$$

Thus, the C_1 and C_2 formally appear to share 2/3 of a pbond [Recall that we are omitting the σ orbitals, so the total bond order would be 1 2/3 including the σ bonds]. We can repeat the same procedure for each CC bond in benzene and we will find the same result: there are 6 equivalent π bonds, each of order 2/3. This gives us great confidence in drawing the Lewis structure we all learned in freshman chemistry:

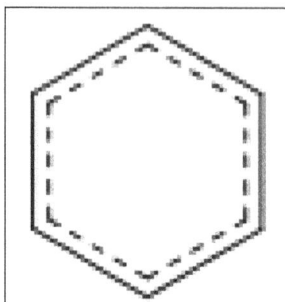

You might have expected this to give a bond order of 1/2 for each CC π-bond rather than 2/3. The extra 1/6 of a bond per carbon comes directly from the aromatic stabilization: because the molecule is more stable than three isolated pbonds by 2β,, this effectively adds another π-bond to the system, which gets distributed equally among all six carbons, resulting in an increased bond order. This effect can be confirmed experimentally, as benzene has slightly shorter CC bonds than non-aromatic conjugated systems, indicating a higher bond order between the carbons.

Just as we can use simple MO theory to describe resonance structures and aromatic stabilization, we can also use it to describe crystal field and ligand field states in transition metal compounds and the sp, sp² and sp³ hybrid orbitals that arise in directional bonding. These results not only mean MO theory is a useful tool – in practice these discoveries have led to MO theory becoming part of the way chemists think about molecules.

References

- Atomic-Orbitals, Electronic-Structure-of-Atoms-and-Molecules, Supplemental-Modules-(Physical-and-Theoretical-Chemistry), Physical-and-Theoretical-Chemistry-Textbook-Maps, Bookshelves: chem.libretexts.org, Retrieved 18 May, 2019

- Clayden, Jonathan; Greeves, Nick; Warren, Stuart; Wothers, Peter (2001). Organic Chemistry (1st ed.). Oxford University Press. Pp. 96–103. ISBN 978-0-19-850346-0

- Definition-of-molecular-orbital-605367: thoughtco.com, Retrieved 19 June, 2019

- Scerri, Eric (7 November 2013). "The trouble with the aufbau principle". Education in Chemistry. Vol. 50 no. 6. Royal Society of Chemistry. Pp. 24–26. Archived from the original on 21 January 2018. Retrieved 12 June 2018

- Huckel, recitations, chem-c2407-archive, chemistry: columbia.edu, Retrieved 20 July, 2019

- Lecture, lecture-notes, physical-chemistry-fall-2007, chemistry, courses: ocw.mit.edu, Retrieved 21 August, 2019

7

Diverse Aspects of Quantum Chemistry

Some of the important concepts of quantum chemistry include Born–Oppenheimer approximation, Zeeman effect, Brillouin's theorem, Koopmans' theorem, Schrödinger equation, etc. This chapter closely examines these key concepts of quantum chemistry to provide an extensive understanding of the subject.

Slater Determinant

Therefore a general method is needed to construct electron wave functions which are completely antisymmetric in all cases. For this purpose, we can use the complete antisymmetry property of determinants. Since interchanging any two rows (or columns) of a determinant changes the sign of the determinant, completely antisymmetric wave functions can be expressed in the form of a determinant.

Let us consider the case of N electrons. Let $\varphi, \ldots, \varphi$ be some arbitrary "one-particle states" of electrons. By a "one-particle state" we mean the spin and space part of a wave function of single electron. So, φ_i might be $\psi_{100}(r)|\uparrow\rangle$, or $\psi_{21-1}(r, \theta, \phi)\frac{1}{\sqrt{2}}(|\uparrow\rangle-|\downarrow\rangle)$. They do not even need to be a product of spin and space parts, so $\frac{1}{\sqrt{3}}\left(\psi_{100}(r)\left|\uparrow\langle+\sqrt{2}\psi_{211}\left|\downarrow\rangle\right)\right.\right.$ is possible as well.

We will also have a short-hand notation for the particle coordinates. The space and spin coordinates of the ith particle will be denoted by the single number i. We haven't been very precise in what we mean by "spin coordinate" of a particle, since it is fairly obvious. So by $\varphi_j(i)$ we mean that the ith particle's space and spin is in the one-particle state φ_j. Since exchange should change both the space coordinate and the spin coordinate, this short-hand notation is quite useful in this case.

For the state where these N electrons are in these N one-particle states, the wavefunction of the whole system can be written as.

$$\Psi(1,\ 2,\ \dots,\ N)\ =\begin{vmatrix}\varphi_1(1) & \varphi_1(2) & \cdots & \varphi_1(N) \\ \varphi_2(1) & \varphi_2(2) & \cdots & \varphi_2(N) \\ \vdots & \vdots & \ddots & \vdots \\ \varphi_N(1) & \varphi_N(2) & \cdots & \varphi_N(N)\end{vmatrix}$$

Where we have ignored the normalization factor. This expression is called as the Slater determinant of N states $\varphi_1, \dots, \varphi_N$. As can be seen, exchanging the coordinates of ith and j^{th} particles' coordinates is equivalent to interchanging i^{th} and j^{th} columns of the determinant. As a result, exchange of any two particles' coordinates changes the sign of the wavefunction of the system Ψ. We have managed to obtain a totally anti-symmetric wavefunction. Although the equation does not show us well, there is a price that we pay for this: Ψ may not be chosen as an eigenstate of some operators (like S^2). For the time being, we will postpone the discussion of the physical meaning of the state Ψ.

We should also note that, since the determinant is also totally antisymmetric rowwise, interchanging the places of any two one-particle states will change the sign of Ψ. One implication of this property is that when you have chosen a set $\varphi_1, \dots, \varphi_N$ such that two states are the same, e.g., $\varphi i = \varphi j$, then the wavefunction of the system is identically zero. This is because, when you interchange the positions of ith and j^{th} states, Ψ should remain the same since the one-particle-state set is unchanged. On the other hand, it should change sign because of the antisymmetry of the determinant. Last two sentences are consistent only if Ψ is identically zero. Since an identically zero wavefunction is not possible in quantum mechanics, we should have distinct one-particle states in the set $\varphi_1, \dots, \varphi_N$. This is the famous "Pauli exclusion principle". No two electrons in a system can be in the same one-particle state. Although we have shown this only for Slater determinants, we can extend this statement for other types of system wavefunctions as well. Note that in the statement "one-particle state" refers to both space and spin parts.

Choosing One-particle States

We have said that if two one-particle states are the same, the determinant is identically zero, since in that case two rows in the determinant will be exactly identical for all possible values of the particle coordinates. There are other properties of the determinants which will enable us to do further simplifications. For example, multiplying a row by a constant and adding to another row will not change the value of a determinant. As a result, for example we can replace the one-particle state φ_1 with $\varphi_1 + a\varphi_2$ and the system's wavefunction Ψ will remain the same. Since, only the system's wavefunction is relevant for the physical properties of the system, we have some freedom in choosing one-particle states.

First of all we can extend the "Pauli exclusion principle" a little bit. If the one-particle state set $\varphi_1, \dots, \varphi_N$ is linearly dependent (that means one state can be written as a linear combination of others, e.g., $\varphi_1 = a\varphi_2 + b\varphi_3$) the Ψ is identically zero. We have to make sure that the set $\varphi_1, \dots, \varphi_N$ is formed by N independent one-particle states. Their being distinct is not enough. Linear independence of the N one-particle states is enough to guarantee that Ψ is not identically zero. Proof of this statement is left to the reader.

There is still some freedom in choosing the one-particle states, and it is convenient to take advantage of this. First of all, we need to explore how much freedom we have. We claim that for a given set of one-particle states $\varphi_1, \dots, \varphi_N$, by taking any linear combinations of this set and forming any other N linearly independent states, say $\varphi_1', \dots, \varphi_N'$, we won't change Ψ. Let us be a little bit more specific. Let us define the new one-particle states φ_1' as:

$$\varphi_i' = \sum_{j=1}^{N} A_{ij} \varphi_j,$$

where A_{ij} are some complex numbers. The new set φ_i' is linearly independent only if the determinant of the matrix A is non-zero. In this case the old set can be written in terms of the new set as well.

$$\varphi_i = \sum_{j=1}^{N} (A^{-1})_{ij} \varphi_j'.$$

Now, what happens when we form the Slater determinant from the new set φ_i'? To see this clearly, we need to re-express Eq. $\varphi_i' = \sum_{j=1}^{N} A_{ij} \varphi_j$, in matrix notation:

$$\begin{bmatrix} \varphi_1' \\ \varphi_2' \\ \vdots \\ \varphi_N' \end{bmatrix} \begin{bmatrix} A_{11} & A_{12} & \cdots & A_{1N} \\ A_{21} & A_{22} & \cdots & A_{2N} \\ \vdots & \vdots & \ddots & \vdots \\ A_{N1} & A_{N2} & \cdots & A_{NN} \end{bmatrix} \begin{bmatrix} \varphi_1 \\ \varphi_2 \\ \vdots \\ \varphi_N \end{bmatrix}$$

It is fairly straightforward to change this relationship to a relationship between Slater determinants. First, in matrix form:

$$\begin{bmatrix} \varphi_1'(1) & \cdots & \varphi_1'(N) \\ \varphi_2'(1) & \cdots & \varphi_2'(N) \\ \vdots & \ddots & \vdots \\ \varphi_N'(1) & \cdots & \varphi_N'(N) \end{bmatrix} = \begin{bmatrix} A_{11} & \cdots & A_{1N} \\ A_{21} & \cdots & A_{2N} \\ \vdots & \ddots & \vdots \\ A_{N1} & \cdots & A_{NN} \end{bmatrix} \begin{bmatrix} \varphi_1 & \cdots & \varphi_1 \\ \varphi_2 & \cdots & \varphi_2 \\ \vdots & \ddots & \vdots \\ \varphi_N & \cdots & \varphi_N \end{bmatrix}$$

Since the determinant of a product of matrices is equal to the product of determinants we have:

$$\Psi' \, 0 \, (1, \, 2, \, \dots, \, N) = \det(A) \Psi(1, \, 2, \, \dots, \, N),$$

where Ψo is the wavefunction of the system prepared by the new set of one particle states $\varphi_1', \dots, \varphi_N'$ and Ψ is the one that is prepared by the old set $\varphi_1', \dots, \varphi_N'$. Therefore the wavefunctions of the system prepared by the two sets differ only by a normalization factor. Since, that means the same expectation values for physical observables (if proper normalization is carried out), we claim that the "same" wavefunction is obtained in both cases.

As a result, we see that in the choice of the one-particle states $\phi_1, \, \dots, \, \phi_N$, neither order nor the particular linear combination is important. As a concrete example, a Slater determinant constructed from the states 1s-spin-up and 1s-spin-down is not different than the one constructed from

1s-spin-x-up and 1s-spin-x-down. There is no physical way of distinguishing the way the determinant is constructed.

Because of this, we choose the one-particle states as an orthonormal set. Gram-Schmidt process enables us to do this for any independent set of N states. As a result we will assume that the relations:

$$\langle \varphi_i \mid \varphi_j \rangle \delta_{ij}, \quad i, j = 1, \ldots, N$$

are satisfied. This orthonormality property that we require will be very useful in evaluating expectation values of observables. Even with that restriction, there is some arbitrariness in the choice of the particular states of the set. We can still take linear combinations:

$$\varphi_i' = \sum_{j=1}^{N} A_{ij} \varphi_j,$$

and if the matrix-A is unitary, the new set will be orthonormal as well. This kind of arbitrariness cannot be eliminated further. At this point, the user should decide which particular linear combinations are useful. So, if the two electron system mentioned above is in a magnetic field along x-axis, then you may want to choose your states as 1s-spin-x-up and 1s-spin-x-down.

Normalization

Before going further we need to normalize the system's wavefunction. We need to keep in mind that in evaluating $\langle \Psi \mid \Psi \rangle$ integrals over particles' coordinates should be unlimited. This is a choice that we have made at the beginning when we were discussing the wavefunctions of identical particles. For the case of a Slater determinant formed by N orthonormal one-particle states, it is useful to have an expanded form of the determinants. Expanding the determinant is easy for $N = 2$ and perhaps for $N = 3$, but for larger N values, it becomes increasingly cumbersome to write the expansion. For this reason we need a shortened way of representing the determinant.

First we note that the determinant will be a sum of terms that are obtained by multiplying N elements of the matrix coming from different rows and different columns. As a result, the first few terms in the expansion of Ψ will be:

$$\Psi(1, 2, \ldots, N) = \varphi_1(1)\varphi_2(2)\varphi_3(3) \ldots \varphi_N(N)$$
$$-\varphi_1(2)\varphi_2(1)\varphi_3(3) \ldots \varphi_N(N) + \ldots$$

There will be $N!$ terms in the expansion which will be formed by terms like:

$$\pm \varphi_{i1}(1)\varphi_{i2}(2) \ldots \varphi_{iN}(N),$$

where $i_1, i_2, \ldots, i_N$ are distinct numbers from the set 1, 2, ..., N. We say that $i_1, i_2, \ldots, N$ is a permutation of $i_1, i_2, \ldots, i_N$. Since there are $N!$ different permutations of an N element set, we have N! terms in the expansion of the determinant. The only problem is to decide whether the term should have a + or − sign.

A property that we should keep in mind is that when we have a permutation and we exchange the places of two labels, the new permutation should have the opposite sign. Since the "identity" permutation 12 . . . N should have + sign, you can find the sign of any permutation starting from the identity permutation. As a definition, a permutation which can be obtained from the identity permutation by an even number of exchanges is called an even permutation and should have + sign. On the other hand, a permutation which can be obtained from the identity permutation by an odd number of exchanges is called an odd permutation and should have − sign

There is a faster way of finding the sign of a permutation which might be useful. When you write down the permuted labels, you count the number of the pairs which are out of order. Or, in other words, for each label in the list you count the number of labels which came before but was larger than the particular label you are considering. If the sum of all of these numbers are even, then the permutation is even and similarly for the odd case. For example in the permutation 143265 of six labels, the sum of the numbers mentioned above is $0 + 0 + 1 + 2 + 0 + 1 = 4$, as a result the permutation is even.

Denoting a permutation by P, its labels by $P_1, P_2, \ldots, PN$ and its sign by $(-)^P$ we can express the Slater determinant as:

$$\Psi(1, 2, \ldots, N) = \sum_P (-)^P \varphi_{P1}(1)\varphi_{P2}(2)\cdots\varphi_{PN}(N).$$

Note that each term in this summation is normalized and any two different terms are orthogonal. This is because the inner product of the term with permutation P and the one with permutation Q is:

$$(-)^P (-)^Q \langle \varphi_{P1}(1)\cdots\varphi_{PN}(N)|\varphi_{Q1}(1)\cdots\varphi_{QN}(N)\rangle =$$
$$(-)^P (-)^Q \delta_{P1,Q1}\cdots\delta_{PN,QN}$$

Since the Slater determinant is a sum of N! orthonormal terms, the inner product of Ψ with itself is simly:

$$\langle \Psi \mid \Psi \rangle = N!.$$

Hence, including the normalization, the Slater determinant of N orthonormal one-particle states can be written as:

$$\Psi(1, 2, \ldots, N)\frac{1}{\sqrt{N!}} = \begin{vmatrix} \varphi_1(1) & \varphi_1(2) & \cdots & \varphi_1(N) \\ \varphi_2(1) & \varphi_2(2) & \cdots & \varphi_2(N) \\ \vdots & \vdots & \ddots & \vdots \\ \varphi_N(1) & \varphi_N(2) & \cdots & \varphi_N(N) \end{vmatrix}$$

This is the form that we will use when evaluating the expectation values.

Expectation values single-particle observables We start with the expectation values of single-particle observables. Consider an arbitrary operator A which expresses a property of only one particle. For example, A might be a component of position (x), of momentum (p_y) or of spin (S_z). We will use the notation A(i) to show that the operator depends on i th particle's properties. So, if A(1) = x_1, then A(2) = x_2, etc. Since the wavefunction of the system Ψ is quite complicated, we need to find a simple expression for the expectation value of A(i).

The first thing that we should note is that the expectation value of A for particle-1 should not be different from the expectation value of A for any other particle. i.e.,

$$\langle A(1) \rangle = \langle \Psi | A(1) | \Psi \rangle = \langle A(2) \rangle = \ldots = \langle A(N) \rangle.$$

This is not a result of the particular properties of the Slater determinants. It is due to the identicalness of the electrons. Same relations should hold for an arbitrary totally antisymmetric wavefunction as well. It just expresses the fact that there is no physical way of distinguishing particles in your system. When you try to calculate the expectation value of $A(1)$, you will find the expression.

$$\langle A(1) \rangle = \frac{1}{N!} \sum_{P,Q} (-)^P (-)^Q \langle \varphi_{P1}(1) | A(1) | \varphi_{Q1}(1) \rangle \delta_{P2,Q2} \cdots \delta_{PN,QN}$$

It is not difficult to see that in the summation the permutations P and Q need to be same if a non-zero result should be found. Nor it is difficult to sum over most of the terms. The final result for the expectation value is:

$$\langle \Psi | A(1) | \Psi \rangle = \frac{1}{N} \sum_{i=1}^{N} \langle \varphi_i(1) | A(1) | \varphi_i \rangle.$$

In the terms within the summation we have expectation values of A in each of the N one-particle states. In here, there is no need to keep the label-1, which denotes space and spin coordinates of the first particle. As a result, we can rewrite the last expression as:

$$\langle \Psi | A(1) | \Psi \rangle = \frac{1}{N} \sum_{i=1}^{N} \langle \varphi_i | A | \varphi_i \rangle.$$

In these kinds of expressions, care is needed to distinguish in which state the expectation value is taken. On the left, we have the expectation value in the state of the system (the physically relevant ones), and on the right we have the expectation values in the one-particle states used to prepare the Slater determinant (physically inaccessible ones).

The last expression can be utilized to calculate the expectation values of single-particle operators. But it has a clear physical meaning as well. It just expresses the fact that the particle-1 sees an equal probability of occupying each of the N states. As a result, property-A of the particle-1 will be the arithmetic average of the properties in each state.

Finally, one is rarely interested in a property of a single particle in an N-particle system. For example, consider the case where $A(1)$ denotes the electric field created by particle-1 at a certain point in space. Obviously, electric field at that certain point will not be created by the first particle only. So, for that physical quantity we should be interested in $A(1) + A(2) + \ldots + A(N)$ as the electric field created by all of the N electrons. This is a general property of systems composed of identical particles. Any physically relevant observables should remain the same under particle exchanges. Although you can form observables (operators) which does not have this property (like $A(1)$), and evaluate their expectation values, they cannot be measured by a physical apparatus. Now the expectation value of $A(1) + A(2) + \ldots + A(N)$ can be expressed as:

$$\langle \Psi \mid \{A(1) + A(2) + \ldots + A(N)\} \mid \Psi \rangle =$$
$$\langle \varphi_1 \mid A \mid \varphi_1 \rangle + \langle \varphi_2 \mid A \mid \varphi_2 + \rangle \ldots + \langle \varphi_N \mid A \mid \varphi_N \rangle$$

So, the sum of a property over particles is equal to the sum of the same property over one-particle states.

Expectation Values of Two-particle Observables

The simplification obtained above is not present for observables depending on two particles. Such a property might be the Coulomb interaction potential between electrons:

$$A(1, 2) = \frac{e^2}{|r_1 - r_2|},$$

(which is usually the case) or it might be operators like $A(1, 2) = x_1 p_{y2} + x_2 p_{y1}$. You should make sure that the operator is symmetric under the exchange of two coordinates since they have to be physically relevant. [In principle you may evaluate the expectation values of the operators like $A(1, 2) = x_1 p_{y2}$, but such operators will never arise in actual problems. You need to extend the expression given below to deal with those kinds of operators.] Now, the expectation value of A(1, 2) can be expressed as:

$$\langle \Psi \mid A(1,2) \mid \Psi \rangle = \frac{1}{N!} \sum_{P,Q} (-)^P (-)^Q \langle \varphi_{P1} \varphi_{P2} \mid A \mid \varphi_{Q1} \varphi_{Q2} \rangle \delta_{P3,Q3} \ldots \delta_{PN,QN}.$$

In this case, if a term in the summation is nonzero, either P and Q should be the same, or they should differ by an exchange of first and second labels. Because of this, the expectation value can be expressed as:

$$\langle \Psi \mid A(1,2) \mid \Psi \rangle = \frac{2}{N(N-1)} \sum_{i<j} \langle \varphi_i \varphi_j \mid A \mid \varphi_i \varphi_j \rangle - \langle \varphi_j \varphi_i \mid A \mid \varphi_i \varphi_j \rangle.$$

Here we needed to use a short-hand notation for the two-particle integrals on the left hand side. This notation is quite widely used. When the one-particle states do not include the spin degree of freedom, the definition is:

$$\langle \phi \psi \mid A \mid \chi \xi \rangle = \iint \phi(r_1)^* \psi(r_2)^* A(1, 2) \chi(r_1) \xi(r_2) dr_1 dr_2.$$

We don't dare to write the right hand side when the spin is included, especially when A(1, 2) includes spin operators. But how the spin part is handled should be obvious. In $\langle \phi \psi \mid A \mid \chi \xi \rangle$, the first states on the bra and ket sides (ϕ and χ) have the same particle's coordinate. And similarly for the second states.

In the first term in the summation of Eq.

$$(\langle \Psi \mid A(1,2) \mid \Psi \rangle = \frac{2}{N(N-1)} \sum_{i<j} \langle \varphi_i \varphi_j \mid A \mid \varphi_i \varphi_j \rangle - \langle \varphi_j \varphi_i \mid A \mid \varphi_i \varphi_j \rangle),$$

the states on the bra and the ket sides are the same and they are in the same order. Thus, this first term is the same thing as what we mean by "expectation value". However, in the second term, $\langle \varphi_j\varphi_i | A | \varphi_i\varphi_j \rangle$, the order of the states are different in the bra and ket sides. For this reason, these terms (and similar terms that will arise later) are called exchange terms. Exchange terms do not have a classical counterpart. To see why, consider the case where A is the Coulomb potential energy of two electrons. In this case, the first terms give (ignoring again the spin parts).

$$\langle \varphi_j\varphi_i | A | \varphi_i\varphi_j \rangle = \iint |\varphi_i(r_1)|^2 |\varphi_j(r_2)|^2 \frac{e^2}{|r_1 - r_2|} dr_1 dr_2 ,$$

which can be interpreted as the expectation value of Coulomb energy when the first particle is in the state φi and the second particle is in the state φj. On the other hand, the exchange term,

$$\langle \varphi_j\varphi_i | A | \varphi_i\varphi_j \rangle = \iint \varphi_i(r_1)^*\varphi_i(r_2)\varphi_j(r_2)^*\varphi_j(r_1) \frac{e^2}{|r_1 - r_2|} dr_1 dr_2 ,$$

is something else. We cannot give a nice physical meaning to it immediately. However, the exchange terms have a contribution to the Coulomb energy of two particles in the N particle system. Where did this term come from? Basic interpretation goes something like this: The Coulomb interaction between the particles can cause the particles to exchange their places. (Is this physically meaningful?) This issue does not arise for distinguishable particles. Only for identical particles such a path is open. Presence of such an exchange processes, then, affects the expectation values of two-particle observables. The meaning of this last interpretation is left to the reader. But, mathematically, exchange terms are present and we have to use them to obtain the physically relevant results.

The Question of Independence

We say that two statistical quantities A and B are uncorrelated if the statistical properties of A and B are independent of each other. (Distribution of A is not dependent on which value we find for B). In terms of averages this implies that $\langle AB \rangle = \langle A \rangle \langle B \rangle$. That property should be valid for any property of A and B, so we should have $\langle f(A)g(B) \rangle = \langle f(A) \langle g(B)$ for all functions f and g. For the case of particles, we say that two particles are independent (or they are uncorrelated) if $\langle A(1)B(2) \rangle = \langle A(1) \rangle \langle g(B) \rangle$ for all observables A and B. This is only possible if the wavefunction of the whole system is a product of the wavefunctions of two particles:

$$\Psi(1,2) = \varphi_1(1)\varphi_2(2)$$

Only in that case we can say that any property of particle-1 can be measured by disregarding particle-2. Any property of particle-1 is independent of what the second particle is doing.

In classical mechanics we will have independent particles if the particles are not interacting either directly or indirectly. (If 1 interacts with 3 and 2 interacts with 3, then 1 and 2 interacts indirectly although there may not be a direct interaction between them). In quantum mechanics, the absence of the interaction is not enough. You also need to have a wavefunction which is uncorrelated.

Independence issue may also arise in the investigation of an isolated system. In that case, we expect that any property of the system is independent of what is going on in the rest of the universe. In such a case, what we mean by independence, in terms of wavefunctions, is, $\Psi_{\text{Universe}} = \psi_{\text{system}} \psi_{\text{rest}}$. As a result, any property of the isolated system, evaluated in the full wavefunction of the universe, Ψ_{Universe}, will be the same as the one calculated in the wavefunction of the system only (ψ_{system}), ignoring the rest of the universe. This is what we have been doing when we were solving the Hydrogen atom problem. In principle, though, we might have correlations between the particular hydrogen atom and the rest of the universe.

Now, for the case of identical fermions, the wavefunctions of many-particle systems cannot be written as a simple product. The best we can do is to write them in Slater determinants. As a result, identical particles can never be independent by the definition given above. This is actually troublesome, since if we want to investigate the behavior of an electron in a particular hydrogen atom, we cannot claim that this electron is behaving independently from what the electrons on the Moon are doing.

It is important that we settle the questions that arise in the paragraph above. When we claim "independence" we have to separate the (direct or indirect) interaction effects from the effects arising from indistinguishability. You can claim that your electron is independent of the other electrons in the Moon because the Coulomb potentials of the Moon's electrons are negligible at the location of your hydrogen atom. This is OK. On the other hand, the most troublesome is the effects caused by identicalness. In that case, it seems that the relevant concept that has to to used is the "exchange effects". Since the probability of your electron exchanging places with one of the electrons on the Moon is small, the "exchange processes" cause a negligible change in what you are measuring.

In any case, we need a separate definition of independence for identical particles. Since the Slater determinant is the simplest expression for the many-body wavefunction of many fermions, we will say that the fermions are independent if their wavefunction is a Slater determinant. There is a good reason for this definition. If the N particles in a system are not interacting, then the Hamiltonian of the system can be written as:

$$H = h(1) + h(2) + \ldots + h(N).$$

Where h is a one-particle hermitian operator that can be called as one-particle Hamiltonian. For the case of N non-interacting electrons in an atom, h is:

$$h(i) = \frac{p_i^2}{2m} - \frac{Ze^2}{r_i},$$

ignoring also the spin-orbit coupling. Now, if φi are the eigenstates of h with eigenvalue $\in_i$, i.e.,

$$h\varphi_i = \in_i \varphi_i \ ,$$

than the complete set of eigenstates of H are the Slater determinants of any N distinct subsets of φ :

$$\Psi_{i_1,i_2,\ldots,i_N} = \frac{1}{\sqrt{N!}} \begin{vmatrix} \varphi_{i_1}(1) & \cdots & \varphi_{i_1}(N) \\ \vdots & \ddots & \vdots \\ \varphi_{i_N}(1) & \cdots & \varphi_{i_N}(N) \end{vmatrix}$$

$$H\Psi_{i_1,i_2,...,i_N} = E_{i_1,i_2,...,i_N} \Psi_{i_1,i_2,...,i_N} = (\epsilon_{i1} + ... + \epsilon_{iN})\Psi_{i_1,i_2,...,i_N}.$$

Since, the eigenfunctions of a noninteracting system of identical particles is of the form of a Slater determinant, and since noninteracting means independent (at least classically), then all Slater determinants describe independent identical particles.

A Slater determinant is the most independent wavefunction possible for identical particles. We obviously do not have $\langle A(1)B(2)\rangle = \langle A(1)\rangle\langle B(2)\rangle$ for Slater determinants. Hence the particles are always correlated. And this is the minimum amount of correlation that you will get for identical particles.

When the particles are interacting, eigenfunctions of the Hamiltonian will not be of the form of Slater determinants. Hence, writing down an expression for the eigenstates of the Hamiltonian will be quite difficult. In those cases, we might take linear combinations of different Slater determinants. Even in this case, there will be too many possible wavefunctions that will make the life difficult for you. As an example, consider the Lithium atom with three electrons. You may want to use only the 1s, 2s and the 2p states to write down a wavefunction which will have a "dependent" electron character. That implies that there are two 1s, two 2s and six 2p states which will be available in forming possible Slater determinants. Three element subsets of these ten one-particle states can be formed in 120 possible ways. Hence if you don't have a way of radically cutting down this number to reasonable levels (for example by using conservation of S_z and L_z) then you will have too many terms to deal with. This is the main difficulty in many-particle problems. There are too many possible states for the wavefunction of the system (which grows exponentially with the size of the system). As a result, you cannot form wavefunctions where the state is sufficiently far away from independent particle case. Even if you do form, you cannot evaluate expectation values.

Born–Oppenheimer Approximation

Born–Oppenheimer (BO) approximation is the assumption that the motion of atomic nuclei and electrons in a molecule can be treated separately. The approach is named after Max Born and J. Robert Oppenheimer who proposed it in 1927, in the early period of quantum mechanics. The approximation is widely used in quantum chemistry to speed up the computation of molecular wavefunctions and other properties for large molecules. There are cases where the assumption of separable motion no longer holds, which make the approximation lose validity (it is said to "break down"), but is then often used as a starting point for more refined methods.

In molecular spectroscopy, using the BO approximation means considering molecular energy as a sum of independent terms, e.g.: $E_{\text{total}} = E_{\text{electronic}} + E_{\text{vibrational}} + E_{\text{rotational}} + E_{\text{nuclear spin}}$. These terms are of different order of magnitude and the nuclear spin energy is so small that it is often omitted. The electronic energies $E_{\text{electronic}}$ consist of kinetic energies, interelectronic repulsions, internuclear repulsions, and electron–nuclear attractions, which are the terms typically included when computing the electronic structure of moleucles.

The benzene molecule consists of 12 nuclei and 42 electrons. The Schrödinger equation, which must be solved to obtain the energy levels and wavefunction of this molecule, is a partial differential eigenvalue equation in the three-dimensional coordinates of the nuclei and electrons, giving 3×12 + 3×42 = 36 nuclear + 126 electronic = 162 variables for the wave function. The computational complexity, i.e. the computational power required to solve an eigenvalue equation, increases faster than the square of the number of coordinates.

When applying the BO approximation, two smaller, consecutive steps can be used: For a given position of the nuclei, the electronic Schrödinger equation is solved, while treating the nuclei as stationary (not "coupled" with the dynamics of the electrons). This corresponding eigenvalue problem then consists only of the 126 electronic coordinates. This electronic computation is then repeated for other possible positions of the nuclei, i.e. deformations of the molecule. For benzene, this could be done using a grid of 36 possible nuclear position coordinates. The electronic energies on this grid are then connected to give a potential energy surface for the nuclei. This potential is then used for a second Schrödinger equation containing only the 36 coordinates of the nuclei.

So, taking the most optimistic estimate for the complexity, instead of a large equation requiring at least $168^2 = 26,244$ hypothetical calculation steps, a series of smaller calculations requiring $N126^2 = N15,876$ (with N being the amount of grid points for the potential) and a very small calculation requiring $36^2 = 1296$ steps can be performed. In practice, the scaling of the problem is larger than n^2 and more approximations are applied in computational chemistry to further reduce the number of variables and dimensions.

The slope of the potential energy surface can be used to simulate Molecular dynamics, using it to express the mean force on the nuclei caused by the electrons and thereby skipping the calculation of the nuclear Schrödinger equation.

The BO approximation recognizes the large difference between the electron mass and the masses of atomic nuclei, and correspondingly the time scales of their motion. Given the same amount of kinetic energy, the nuclei move much more slowly than the electrons. In mathematical terms, the BO approximation consists of expressing the wavefunction (Ψ_{total}) of a molecule as the product of an electronic wavefunction and a nuclear (vibrational, rotational) wavefunction. $\emptyset_{total} = \psi_{electronic}\psi_{nuclear}$. This enables a separation of the Hamiltonian operator into electronic and nuclear terms, where cross-terms between electrons and nuclei are neglected, so that the two smaller and decoupled systems can be solved more efficiently.

In the first step the nuclear kinetic energy is neglected, that is, the corresponding operator T_n is subtracted from the total molecular Hamiltonian. In the remaining electronic Hamiltonian H_e the nuclear positions are no longer variable, but are constant parameters (they enter the equation "parametrically"). The electron–nucleus interactions are not removed, i.e., the electrons still "feel" the Coulomb potential of the nuclei clamped at certain positions in space. This first step of the BO approximation is therefore often referred to as the clamped-nuclei approximation.

The electronic Schrödinger equation:

$$H_e(\mathbf{r},\mathbf{R})\chi(\mathbf{r},\mathbf{R}) = E_e\chi(\mathbf{r},\mathbf{R}),$$

is solved approximately The quantity r stands for all electronic coordinates and R for all nuclear coordinates. The electronic energy eigenvalue E_e depends on the chosen positions R of the nuclei.

Varying these positions R in small steps and repeatedly solving the electronic Schrödinger equation, one obtains E_e as a function of R. This is the potential energy surface (PES): $E_e(R)$. Because this procedure of recomputing the electronic wave functions as a function of an infinitesimally changing nuclear geometry is reminiscent of the conditions for the adiabatic theorem, this manner of obtaining a PES is often referred to as the adiabatic approximation and the PES itself is called an adiabatic surface.

In the second step of the BO approximation the nuclear kinetic energy T_n (containing partial derivatives with respect to the components of R) is reintroduced, and the Schrödinger equation for the nuclear motion:

$$[T_n + E_e(\mathbf{R})]\phi(\mathbf{R}) = E\phi(\mathbf{R})$$,

is solved. This second step of the BO approximation involves separation of vibrational, translational, and rotational motions. This can be achieved by application of the Eckart conditions. The eigenvalue E is the total energy of the molecule, including contributions from electrons, nuclear vibrations, and overall rotation and translation of the molecule. In accord with the Hellmann-Feynman theorem, the nuclear potential is taken to be an average over electron configurations of the sum of the electron–nuclear and internuclear electric potentials.

Derivation

It will be discussed how the BO approximation may be derived and under which conditions it is applicable. At the same time we will show how the BO approximation may be improved by including vibronic coupling. To that end the second step of the BO approximation is generalized to a set of coupled eigenvalue equations depending on nuclear coordinates only. Off-diagonal elements in these equations are shown to be nuclear kinetic energy terms.

It will be shown that the BO approximation can be trusted whenever the PESs, obtained from the solution of the electronic Schrödinger equation, are well separated:

$$E_0(\mathbf{R}) \ll E_1(\mathbf{R}) \ll E_2(\mathbf{R}) \ll \cdots \text{ for all } \mathbf{R}.$$

We start from the *exact* non-relativistic, time-independent molecular Hamiltonian:

$$H = H_e + T_n$$,

with,

$$H_e = -\sum_i \frac{1}{2}\nabla_i^2 - \sum_{i,A} \frac{Z_A}{r_{iA}} + \sum_{i>j} \frac{1}{r_{ij}} + \sum_{B>A} \frac{Z_A Z_B}{R_{AB}} \quad \text{and} \quad T_n = -\sum_A \frac{1}{2M_A}\nabla_A^2.$$

The position vectors $\mathbf{r} \equiv \{\mathbf{r}_i\}$ of the electrons and the position vectors $\mathbf{R} \equiv \{\mathbf{R}_A = (R_{Axy}, R_{Ayz}, R_{Azx})\}$ of the nuclei are with respect to a Cartesian inertial frame. Distances between particles are written as $r_{iA} \equiv |\mathbf{r}_i - \mathbf{R}_A|$ (distance between electron i and nucleus A) and similar definitions hold for r_{ij} and R_{AB}.

We assume that the molecule is in a homogeneous (no external force) and isotropic (no external torque) space. The only interactions are the two-body Coulomb interactions among the electrons and nuclei. The Hamiltonian is expressed in atomic units, so that we do not see Planck's constant, the dielectric constant of the vacuum, electronic charge, or electronic mass in this formula. The only constants explicitly entering the formula are Z_A and M_A – the atomic number and mass of nucleus A.

It is useful to introduce the total nuclear momentum and to rewrite the nuclear kinetic energy operator as follows:

$$T_n = \sum_A \sum_{\alpha=x,y,z} \frac{P_{A\alpha}P_{A\alpha}}{2M_A} \quad \text{with} \quad P_{A\alpha} = -i\frac{\partial}{\partial R_{A\alpha}}.$$

Suppose we have K electronic eigenfunctions $\chi_k(\mathbf{r};\mathbf{R})$ of H_e, that is, we have solved:

$$H_e \chi_k(\mathbf{r};\mathbf{R}) = E_k(\mathbf{R})\chi_k(\mathbf{r};\mathbf{R}) \quad \text{for} \quad k=1,\dots,K.$$

The electronic wave functions χ_k will be taken to be real, which is possible when there are no magnetic or spin interactions. The parametric dependence of the functions χ_k on the nuclear co-ordinates is indicated by the symbol after the semicolon. This indicates that, although χ_k is a re-al-valued function of r, its functional form depends on R.

For example, in the molecular-orbital-linear-combination-of-atomic-orbitals (LCAO-MO) approx-imation, R is a molecular orbital (MO) given as a linear expansion of atomic orbitals (AOs). An AO depends visibly on the coordinates of an electron, but the nuclear coordinates are not explicit in the MO. However, upon change of geometry, i.e., change of R, the LCAO coefficients obtain differ-ent values and we see corresponding changes in the functional form of the MO χ_k.

We will assume that the parametric dependence is continuous and differentiable, so that it is meaningful to consider:

$$P_{A\alpha}\chi_k(\mathbf{r};\mathbf{R}) = -i\frac{\partial \chi_k(\mathbf{r};\mathbf{R})}{\partial R_{A\alpha}} \quad \text{for} \quad \alpha=x,y,z,$$

which in general will not be zero.

The total wave function $\Psi(\mathbf{R},\mathbf{r})$ is expanded in terms of $\chi_k(\mathbf{r};\mathbf{R})$:

$$\Psi(\mathbf{R},\mathbf{r}) = \sum_{k=1}^{K} \chi_k(\mathbf{r};\mathbf{R})\phi_k(\mathbf{R}),$$

with,

$$\langle \chi_{k'}(\mathbf{r};\mathbf{R}) \mid \chi_k(\mathbf{r};\mathbf{R})\rangle_{(\mathbf{r})} = \delta_{k'k},$$

and where the subscript r indicates that the integration, implied by the braket notation, is over electronic coordinates only. By definition, the matrix with general element:

$$\left(\mathbb{H}_e(\mathbf{R})\right)_{k'k} \equiv \langle \chi_{k'}(\mathbf{r};\mathbf{R}) \mid H_e \mid \chi_k(\mathbf{r};\mathbf{R})\rangle_{(\mathbf{r})} = \delta_{k'k}E_k(\mathbf{R})$$
,

is diagonal. After multiplication by the real function $\chi_{k'}(\mathbf{r};\mathbf{R})$ from the left and integration over the electronic coordinates $\mathbf{r}$ the total Schrödinger equation:

$$H\Psi(\mathbf{R},\mathbf{r}) = E\Psi(\mathbf{R},\mathbf{r}),$$

is turned into a set of K coupled eigenvalue equations depending on nuclear coordinates only:

$$[\mathbb{H}_n(\mathbf{R}) + \mathbb{H}_e(\mathbf{R})]\phi(\mathbf{R}) = E\phi(\mathbf{R}).$$

The column vector $\phi(\mathbf{R})$ has elements $\phi_k(\mathbf{R})$, $k = 1,\dots,K$. The matrix $\mathbb{H}_e(\mathbf{R})$ is diagonal, and the nuclear Hamilton matrix is non-diagonal; its off-diagonal (vibronic coupling) terms $(\mathbb{H}_n(\mathbf{R}))_{k'k}$ are further discussed below. The vibronic coupling in this approach is through nuclear kinetic energy terms.

Solution of these coupled equations gives an approximation for energy and wavefunction that goes beyond the Born–Oppenheimer approximation. Unfortunately, the off-diagonal kinetic energy terms are usually difficult to handle. This is why often a diabatic transformation is applied, which retains part of the nuclear kinetic energy terms on the diagonal, removes the kinetic energy terms from the off-diagonal and creates coupling terms between the adiabatic PESs on the off-diagonal.

If we can neglect the off-diagonal elements the equations will uncouple and simplify drastically. In order to show when this neglect is justified, we suppress the coordinates in the notation and write, by applying the Leibniz rule for differentiation, the matrix elements of T_n as:

$$H_n(\mathbf{R})_{k'k} \equiv (\mathbb{H}_n(\mathbf{R}))_{k'k} = \delta_{k'k}T_n - \sum_{A,\alpha}\frac{1}{M_A}\langle \chi_{k'} \mid P_{A\alpha} \mid \chi_k \rangle_{(\mathbf{r})} P_{A\alpha} + \langle \chi_{k'} \mid T_n \mid \chi_k \rangle_{(\mathbf{r})}.$$

The diagonal $(k' = k)$ matrix elements $\langle \chi_k \mid P_{A\alpha} \mid \chi_k \rangle_{(\mathbf{r})}$ of the operator $P_{A\alpha}$ vanish, because we assume time-reversal invariant, so χ_k can be chosen to be always real. The off-diagonal matrix elements satisfy:

$$\langle \chi_{k'} \mid P_{A\alpha} \mid \chi_k \rangle_{(\mathbf{r})} = \frac{\langle \chi_{k'} \mid [P_{A\alpha},H_e] \mid \chi_k \rangle_{(\mathbf{r})}}{E_k(\mathbf{R}) - E_{k'}(\mathbf{R})}.$$

The matrix element in the numerator is:

$$\langle \chi_{k'} \mid [P_{A\alpha},H_e] \mid \chi_k \rangle_{(\mathbf{r})} = iZ_A\sum_i \left\langle \chi_{k'} \left| \frac{(\mathbf{r}_{iA})_\alpha}{r_{iA}^3} \right| \chi_k \right\rangle_{(\mathbf{r})} \quad \text{with} \quad \mathbf{r}_{iA} \equiv \mathbf{r}_i - \mathbf{R}_A.$$

The matrix element of the one-electron operator appearing on the right side is finite.

When the two surfaces come close, $E_k(\mathbf{R}) \approx E_{k'}(\mathbf{R})$, the nuclear momentum coupling term becomes large and is no longer negligible. This is the case where the BO approximation breaks down, and a coupled set of nuclear motion equations must be considered instead of the one equation appearing in the second step of the BO approximation.

Conversely, if all surfaces are well separated, all off-diagonal terms can be neglected, and hence the whole matrix of P_α^A is effectively zero. The third term on the right side of the expression for the matrix element of T_n (the *Born–Oppenheimer diagonal correction*) can approximately be written as the matrix of P_α^A frid and, accordingly, is then negligible also. Only the first (diagonal) kinetic energy term in this equation survives in the case of well separated surfaces, and a diagonal, uncoupled, set of nuclear motion equations results:

$$[T_n + E_k(\mathbf{R})]\phi_k(\mathbf{R}) = E\phi_k(\mathbf{R}) \quad \text{for} \quad k = 1, \ldots, K,$$

which are the normal second step of the BO equations.

We reiterate that when two or more potential energy surfaces approach each other, or even cross, the Born–Oppenheimer approximation breaks down, and one must fall back on the coupled equations. Usually one invokes then the diabatic approximation.

The Born–Oppenheimer Approximation with the Correct Symmetry

To include the correct symmetry within the Born–Oppenheimer (BO) approximation, a molecular system presented in terms of (mass-dependent) nuclear coordinates $\mathbf{q}$ and formed by the two lowest BO adiabatic potential energy surfaces (PES) $u_1(\mathbf{q})$ and $u_2(\mathbf{q})$ is considered. To ensure the validity of the BO approximation, the energy E of the system is assumed to be low enough so that $u_1(\mathbf{q})$ becomes a closed PES in the region of interest, with the exception of sporadic infinitesimal sites surrounding degeneracy points formed by and $u_2(\mathbf{q})$ (designated as $(1, 2)$ degeneracy points).

The starting point is the nuclear adiabatic BO (matrix) equation written in the form:

$$\frac{\hbar^2}{2m}(\nabla + \tau)^2 \Psi + (\mathbf{u} - E)\Psi = 0,$$

where $\emptyset(\mathbf{q})$ is a column vector containing the unknown nuclear wave functions $\psi_k(\mathbf{q})$, $\mathbf{u}(\mathbf{q})$ is a diagonal matrix containing the corresponding adiabatic potential energy surfaces $u_k(\mathbf{q})$, m is the reduced mass of the nuclei, E is the total energy of the system, ∇ is the gradient operator with respect to the nuclear coordinates $\mathbf{q}$, and $\tau(\mathbf{q})$ is a matrix containing the vectorial non-adiabatic coupling terms (NACT):

$$\tau_{jk} = \langle \zeta_j | \nabla \zeta_k \rangle.$$

Here $|\zeta_n\rangle$ are eigenfunctions of the electronic Hamiltonian assumed to form a complete Hilbert space in the given region in configuration space.

To study the scattering process taking place on the two lowest surfaces, one extracts from the above BO equation the two corresponding equations:

$$-\frac{\hbar^2}{2m}\nabla^2\psi_1 + (\tilde{u}_1 - E)\psi_1 - \frac{\hbar^2}{2m}[2\tau_{12}\nabla + \nabla\tau_{12}]\psi_2 = 0,$$

$$-\frac{\hbar^2}{2m}\nabla^2\psi_2 + (\tilde{u}_2 - E)\psi_2 + \frac{\hbar^2}{2m}[2\tau_{12}\nabla + \nabla\tau_{12}]\psi_1 = 0,$$

where $\tilde{u}_k(\mathbf{q}) = u_k(\mathbf{q}) + (\hbar^2 / 2m)\tau_{12}^2 (k = 1, 2)$ $\tau_{12} = \tau_{12}(\mathbf{q})$ is the (vectorial) NACT responsible for the coupling between $u_1(\mathbf{q})$ and $u_2(\mathbf{q})$.

Next a new function is introduced:

$$\chi = \psi_1 + i\psi_2,$$

and the corresponding rearrangements are made:

1. Multiplying the second equation by i and combining it with the first equation yields the (complex) equation:

$$-\frac{\hbar^2}{2m}\nabla^2\chi + (\tilde{u}_1 - E)\chi + i\frac{\hbar^2}{2m}[2\tau_{12}\nabla + \nabla\tau_{12}]\chi + i(u_1 - u_2)\psi_2 = 0.$$

2. The last term in this equation can be deleted for the following reasons: At those points where $u_2(\mathbf{q})$ is classically closed, $\psi_2(\mathbf{q}) \sim 0$ by definition, and at those points where $u_2(\mathbf{q})$ becomes classically allowed (which happens at the vicinity of the $(1, 2)$ degeneracy points) this implies that: $u_1(\mathbf{q}) \sim u_2(\mathbf{q})$ Or $u_1(\mathbf{q}) - u_2(\mathbf{q}) \sim 0$ Consequently, the last term is, indeed, negligibly small at every point in the region of interest, and the equation simplifies to become:

$$-\frac{\hbar^2}{2m}\nabla^2\chi + (\tilde{u}_1 - E)\chi + i\frac{\hbar^2}{2m}[2\tau_{12}\nabla + \nabla\tau_{12}]\chi = 0.$$

In order for this equation to yield a solution with the correct symmetry, it is suggested to apply a perturbation approach based on an elastic potential $u_1(\mathbf{q})$, which coincides with $u_0(\mathbf{q})$ at the asymptotic region.

The equation with an elastic potential can be solved, in a straightforward manner, by substitution. Thus, if χ_0 is the solution of this equation, it is presented as:

$$\chi_0(\mathbf{q} \mid \Gamma) = \xi_0(\mathbf{q}) \exp\left[-i\int_\Gamma d\mathbf{q}' \cdot \tau(\mathbf{q}' \mid \Gamma)\right],$$

where $\tilde{A}$ is an arbitrary contour, and the exponential function contains the relevant symmetry as created while moving along $\tilde{A}$.

The function $\xi_0(\mathbf{q})$ can be shown to be a solution of the (unperturbed/elastic) equation,

$$-\frac{\ }{\ }\nabla\ \xi_0 + (u_0 - E)\xi_0 = 0.$$

Having $\chi_0(\mathbf{q} \mid \Gamma)$, the full solution of the above decoupled equation takes the form:

$$\chi(\mathbf{q} \mid \Gamma) = \chi_0(\mathbf{q} \mid \Gamma) + \eta(\mathbf{q} \mid \Gamma),$$

where $\eta(q \mid \Gamma)$ satisfies the resulting inhomogeneous equation:

$$-\frac{\hbar^2}{2m}\nabla^2\eta + (\tilde{u}_1 - E)\eta + i\frac{\hbar^2}{2m}[2\tau_{12}\nabla + \nabla\tau_{12}]\eta = (u_1 - u_0)\chi_0.$$

In this equation the inhomogeneity ensures the symmetry for the perturbed part of the solution along any contour and therefore for the solution in the required region in configuration space.

The relevance of the present approach was demonstrated while studying a two-arrangement-channel model (containing one inelastic channel and one reactive channel) for which the two adiabatic states were coupled by a Jahn–Teller conical intersection. A nice fit between the symmetry-preserved single-state treatment and the corresponding two-state treatment was obtained. This applies in particular to the reactive state-to-state probabilities, for which the ordinary BO approximation led to erroneous results, whereas the symmetry-preserving BO approximation produced the accurate results, as they followed from solving the two coupled equations.

The Zeeman Effect

The Zeeman effect named after the Dutch physicist Pieter Zeeman, is the effect of splitting of a spectral line into several components in the presence of a static magnetic field. It is analogous to the Stark effect, the splitting of a spectral line into several components in the presence of an electric field. Also similar to the Stark effect, transitions between different components have, in general, different intensities, with some being entirely forbidden (in the dipole approximation), as governed by the selection rules.

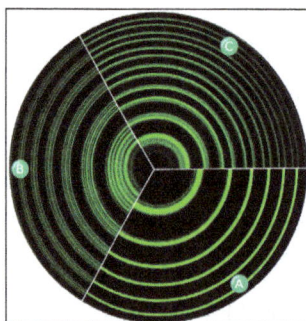

The spectral lines of mercury vapor lamp at wavelength 546.1 nm, showing anomalous Zeeman effect. (A) Without magnetic field. (B) With magnetic field, spectral lines split as transverse Zeeman effect. (C) With magnetic field, split as longitudinal Zeeman effect. The spectral lines were obtained using a Fabry–Pérot interferometer.

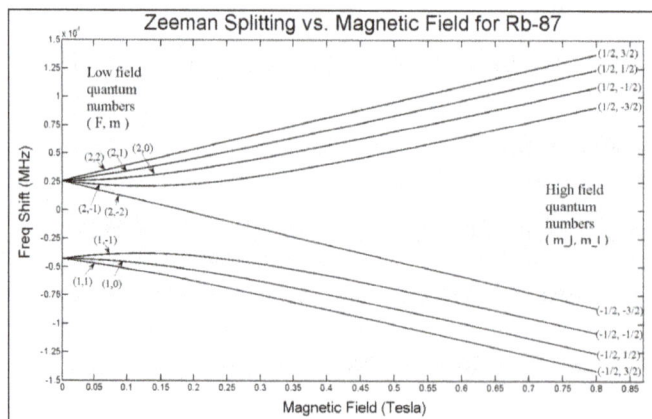

Zeeman splitting of the 5s level of ^{87}Rb, including fine structure and hyperfine structure splitting. Here $F = J + I$, where I is the nuclear spin (for ^{87}Rb, $I = \frac{3}{2}$).

Since the distance between the Zeeman sub-levels is a function of magnetic field strength, this effect can be used to measure magnetic field strength, e.g. that of the Sun and other stars or in laboratory plasmas. The Zeeman effect is very important in applications such as nuclear magnetic resonance spectroscopy, electron spin resonance spectroscopy, magnetic resonance imaging (MRI) and Mössbauer spectroscopy. It may also be utilized to improve accuracy in atomic absorption spectroscopy. A theory about the magnetic sense of birds assumes that a protein in the retina is changed due to the Zeeman effect.

When the spectral lines are absorption lines, the effect is called inverse Zeeman effect.

Nomenclature

Historically, one distinguishes between the normal and an anomalous Zeeman effect (discovered by Thomas Preston in Dublin, Ireland). The anomalous effect appears on transitions where the net spin of the electrons is an odd half-integer, so that the number of Zeeman sub-levels is even. It was called "anomalous" because the electron spin had not yet been discovered, and so there was no good explanation for it at the time that Zeeman observed the effect.

At higher magnetic field strength the effect ceases to be linear. At even-higher field strength, when the strength of the external field is comparable to the strength of the atom's internal field, electron coupling is disturbed and the spectral lines rearrange. This is called the Paschen-Back effect.

In the modern scientific literature, these terms are rarely used, with a tendency to use just the "Zeeman effect".

Theoretical Presentation

The total Hamiltonian of an atom in a magnetic field is:

$$H = H_0 + V_M,$$

where H_o is the unperturbed Hamiltonian of the atom, and V_M is the perturbation due to the magnetic field:

$$V_M = -\vec{\mu} \cdot \vec{B},$$

where $\vec{\mu}$ is the magnetic moment of the atom. The magnetic moment consists of the electronic and nuclear parts; however, the latter is many orders of magnitude smaller and will be neglected here. Therefore,

$$\vec{\mu} \approx -\frac{\mu_B g \vec{J}}{\hbar},$$

where μB is the Bohr magneton, $\vec{J}$ is the total electronic angular momentum, and g is the Landé g-factor. A more accurate approach is to take into account that the operator of the magnetic moment of an electron is a sum of the contributions of the orbital angular momentum $\vec{L}$ and the spin angular momentum $\vec{S}$, with each multiplied by the appropriate gyromagnetic ratio:

$$\vec{\mu} = -\frac{\mu_B (g_l \vec{L} + g_s \vec{S})}{\hbar},$$

where $g_l = 1$ and $g_s \approx 2.0023192$ (the latter is called the anomalous gyromagnetic ratio; the deviation of the value from 2 is due to the effects of quantum electrodynamics). In the case of the LS coupling, one can sum over all electrons in the atom:

$$g\vec{J} = \left\langle \sum_i (g_l \vec{l}_i + g_s \vec{s}_i) \right\rangle = \left\langle (g_l \vec{L} + g_s \vec{S}) \right\rangle,$$

where $\vec{L}$ and $\vec{S}$, are the total orbital momentum and spin of the atom, and averaging is done over a state with a given value of the total angular momentum.

If the interaction term V_M is small (less than the fine structure), it can be treated as a perturbation; this is the Zeeman effect proper. In the Paschen–Back effect, V_M exceeds the LS coupling significantly (but is still small compared to H_O). In ultra-strong magnetic fields, the magnetic-field interaction may exceed H_0, in which case the atom can no longer exist in its normal meaning, and one talks about Landau levels instead. There are intermediate cases which are more complex than these limit cases.

Weak Field (Zeeman Effect)

If the spin-orbit interaction dominates over the effect of the external magnetic field, $\vec{L}$ and $\vec{S}$, are not separately conserved, only the total angular momentum $\vec{J} = \vec{L} + \vec{S}$ is. The spin and orbital angular momentum vectors can be thought of as precessing about the (fixed) total angular momentum vector $\vec{J}$. The averaged spin vector is then the projection of the spin onto the direction of:

$$\vec{S}_{avg} = \frac{(\vec{S} \cdot \vec{J})}{J^2} \vec{J}$$

and for the (time-)"averaged" orbital vector:

$$\vec{L}_{avg} = \frac{(\vec{L} \cdot \vec{J})}{J^2} \vec{J}.$$

Thus,

$$\langle V_M \rangle = \frac{\mu_B}{\hbar} \vec{J} \left(g_L \frac{\vec{L} \cdot \vec{J}}{J^2} + g_S \frac{\vec{S} \cdot \vec{J}}{J^2} \right) \cdot \vec{B}.$$

Using $\vec{L} = \vec{J} - \vec{S}$ and squaring both sides, we get:

$$\vec{S} \cdot \vec{J} = \frac{1}{2}(J^2 + S^2 - L^2) = \frac{\hbar^2}{2}[j(j+1) - l(l+1) + s(s+1)],$$

and: using $\vec{S} = \vec{J} - \vec{L}$ and squaring both sides, we get:

$$\vec{L} \cdot \vec{J} = \frac{1}{2}(J^2 - S^2 + L^2) = \frac{\hbar^2}{2}[j(j+1) + l(l+1) - s(s+1)].$$

Combining everything and taking $J_z = \hbar m_j$, we obtain the magnetic potential energy of the atom in the applied external magnetic field,

$$V_M = \mu_B B m_j \left[g_L \frac{j(j+1)+l(l+1)-s(s+1)}{2j(j+1)} + g_S \frac{j(j+1)-l(l+1)+s(s+1)}{2j(j+1)} \right]$$

$$= \mu_B B m_j \left[1+(g_S -1)\frac{j(j+1)-l(l+1)+s(s+1)}{2j(j+1)} \right],$$

$$= \mu_B B m_j g_j$$

where the quantity in square brackets is the Landé g-factor g_j of the atom ($g_L = 1$ and $g_S \approx 2$) and m_j is the z-component of the total angular momentum. For a single electron above filled shells $s = 1/2$ and $j = l \pm s$, the Landé g-factor can be simplified into:

$$g_j = 1 \pm \frac{g_S - 1}{2l+1}$$

Taking V_M to be the perturbation, the Zeeman correction to the energy is:

$$E_Z^{(1)} = \langle nljm_j | H'Z | nljm_j \rangle = \langle V_M \rangle_\Psi = \mu_B g_J B_{ext} m_j$$

Lyman Alpha Transition in Hydrogen

The Lyman alpha transition in hydrogen in the presence of the spin-orbit interaction involves the transitions:

$$2P_{1/2} \rightarrow 1S_{1/2} \text{ and } 2P_{3/2} \rightarrow 1S_{1/2}.$$

In the presence of an external magnetic field, the weak-field Zeeman effect splits the $1S_{1/2}$ and $2P_{1/2}$ levels into 2 states each ($m_j = 1/2, -1/2$) and the $2P_{3/2}$ level into 4 states ($m_j = 3/2, 1/2, -1/2, -3/2$).). The Landé g-factors for the three levels are:

$g_J = 2$ for $1S_{1/2}$ (j=1/2, l=0),

$g_J = 2/3$ for $2P_{1/2}$ (j=1/2, l=1),

$g_J = 4/3$ for $2P_{3/2}$ (j=3/2, l=1).

Note in particular that the size of the energy splitting is different for the different orbitals, because the g_J values are different. On the left, fine structure splitting is depicted. This splitting occurs even in the absence of a magnetic field, as it is due to spin-orbit coupling. Depicted on the right is the additional Zeeman splitting, which occurs in the presence of magnetic fields.

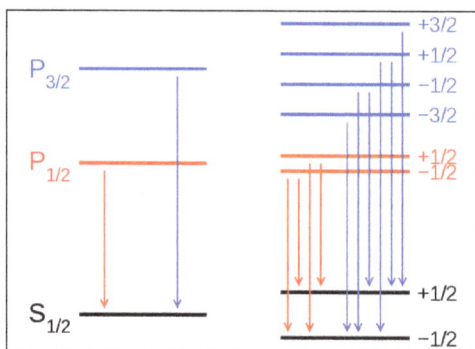

Possible Transitions in the Weak Zeeman Effect		
Initial State $(n=2, l=1)$ $\left\|j, m_j\right\rangle$	Initial Energy Perturbation	Final State $(n=1, l=0)$ $\left\|j, m_j\right\rangle$
$\left\|\dfrac{3}{2}, \pm\dfrac{3}{2}\right\rangle \pm 2\mu_B B_z$		$\left\|\dfrac{1}{2}, \pm\dfrac{1}{2}\right\rangle$
$\left\|\dfrac{3}{2}, \dfrac{1}{2}\right\rangle + \dfrac{2}{3}\mu_B B_z$		$\left\|\dfrac{1}{2}, \pm\dfrac{1}{2}\right\rangle$
$+\dfrac{1}{3}\mu_B B_z$	$+\dfrac{1}{3}\mu_B B_z$	$\left\|\dfrac{1}{2}, \pm\dfrac{1}{2}\right\rangle$
$-\dfrac{1}{3}\mu_B B_z$	$-\dfrac{1}{3}\mu_B B_z$	$\left\|\dfrac{1}{2}, \pm\dfrac{1}{2}\right\rangle$
$\left\|\dfrac{3}{2}, -\dfrac{1}{2}\right\rangle -\dfrac{2}{3}\mu_B B_z$		$\left\|\dfrac{1}{2}, \pm\dfrac{1}{2}\right\rangle$

Paschen–Back Effect

The Paschen–Back effect is the splitting of atomic energy levels in the presence of a strong magnetic field. This occurs when an external magnetic field is sufficiently strong to disrupt the coupling between orbital () and spin () angular momenta. This effect is the strong-field limit of the Zeeman effect. When, the two effects are equivalent. The effect was named after the German physicists Friedrich Paschen and Ernst E. A. Back.

When the magnetic field perturbation significantly exceeds the spin-orbit interaction, one can safely assume. This allows the expectation values of L_z and S_z to be easily evaluated for a state $|\psi\rangle$. The energies are simply:

$$E_z = \left\langle \psi \left| H_0 + \frac{B_z \mu_B}{\hbar}(L_z + g_s S_z) \right| \psi \right\rangle = E_0 + B_z \mu_B (m_l + g_s m_s).$$

The above may be read as implying that the LS-coupling is completely broken by the external field. However m_l and m_s are still "good" quantum numbers. Together with the selection rules for an electric dipole transition, i.e.,

$\Delta s = 0, \Delta m_s = 0, \Delta l = \pm 1, \Delta m_l = 0, \pm 1$ this allows to ignore the spin degree of freedom altogether. As a result, only three spectral lines will be visible, corresponding to the $\Delta m_l = 0, \pm 1$ selection rule. The splitting $\Delta E = B\mu_B \Delta m_l$ is *independent* of the unperturbed energies and electronic configurations of the levels being considered. In general (if $(s \neq 0)$), these three components are actually groups of several transitions each, due to the residual spin-orbit coupling.

In general, one must now add spin-orbit coupling and relativistic corrections (which are of the same order, known as 'fine structure') as a perturbation to these 'unperturbed' levels. First order perturbation theory with these fine-structure corrections yields the following formula for the Hydrogen atom in the Paschen–Back limit:

$$E_{z+fs} = E_z + \frac{m_e c^2 \alpha^4}{2n^3} \left\{ \frac{3}{4n} - \left[\frac{l(l+1) - m_l m_s}{l(l+1/2)(l+1)} \right] \right\}.$$

Possible Lyman-Alpha Transitions in the Strong Effect		
Initial State $(n=2, l=1)$ $\|m_l, m_s\rangle$	Initial Energy Perturbation	Final State $(n=1, l=0)$ $\|m_l, m_s\rangle$
$\left\|1, \frac{1}{2}\right\rangle$	$\pm 2\mu_B B_z$	$\left\|0, -\frac{1}{2}\right\rangle$
$\left\|0, \frac{1}{2}\right\rangle$	$+\mu_B B_z$	$\left\|0, \frac{1}{2}\right\rangle$
$\left\|1, -\frac{1}{2}\right\rangle$	o	$\left\|0, \frac{1}{2}\right\rangle$
$\left\|-1, \frac{1}{2}\right\rangle$	o	$\left\|0, -\frac{1}{2}\right\rangle$
$\left\|0, -\frac{1}{2}\right\rangle$	$-\mu_B B_z$	$\left\|0, -\frac{1}{2}\right\rangle$
$\left\|-1, -\frac{1}{2}\right\rangle$	$-2\mu_B B_z$	$\left\|0, -\frac{1}{2}\right\rangle$

Intermediate field for j = 1/2

In the magnetic dipole approximation, the Hamiltonian which includes both the hyperfine and Zeeman interactions is:

$$H = hA\vec{I}\cdot\vec{J} + (\mu_B g_J \vec{J} + \mu_N g_I \vec{I})\cdot\vec{B}$$
$$H = hA\vec{I}\cdot\vec{J} + (\mu_B g_J \vec{J} + \mu_N g_I \vec{I})\cdot\vec{B},$$

where A is the hyperfine splitting (in Hz) at zero applied magnetic field, μB and μN are the Bohr magneton and nuclear magneton respectively, $\vec{J}$ and $\vec{I}$ are the electron and nuclear angular momentum operators and g j is the Landé g-factor:

$$g_J = g_L \frac{J(J+1)+L(L+1)-S(S+1)}{2J(J+1)} + g_S \frac{J(J+1)-L(L+1)+S(S+1)}{2J(J+1)}.$$

In the case of weak magnetic fields, the Zeeman interaction can be treated as a perturbation to the $|F,m_f\rangle$ basis. In the high field regime, the magnetic field becomes so strong that the Zeeman effect will dominate, and one must use a more complete basis of $|I,J,m_I,m_J\rangle$ or just $|m_I,m_J\rangle$ since I and J will be constant within a given level.

To get the complete picture, including intermediate field strengths, we must consider eigenstates which are superpositions of the $|F,m_F\rangle$ and $|m_I,m_J\rangle$ basis states. For $J=1/2$,, the Hamiltonian can be solved analytically, resulting in the Breit-Rabi formula. Notably, the electric quadrupole interaction is zero for $L=0$ $J=1/2$)), so this formula is fairly accurate.

To solve this system, we note that at all times, the total angular momentum projection $m_F = m_J + m_I$ will be conserved. Furthermore, since $J=1/2$ between states M_J will change between only $\pm 1/2$. . Therefore, we can define a good basis as:

$$|\pm\rangle \equiv |m_J = \pm 1/2, m_I = m_F \mp 1/2\rangle$$

We now utilize quantum mechanical ladder operators, which are defined for a general angular momentum operator L as:

$$L_\pm \equiv L_x \pm iL_y$$

These ladder operators have the property:

$$L_\pm |L,m_L\rangle = \sqrt{(L\mp m_L)(L\pm m_L+1)}\,|L,m_L\pm 1\rangle,$$

as long as M_L lies in the range $L,...,L$ (otherwise, they return zero). Using ladder operators $J_\pm$ and $I\pm$ We can rewrite the Hamiltonian as:

$$H = hAI_zJ_z + \frac{hA}{2}(J_+I_- + J_-I_+) + \mu_B Bg_J J_z + \mu_N Bg_I I_z$$

Now we can determine the matrix elements of the Hamiltonian:

$$\langle\pm|H|\pm\rangle = -\frac{1}{4}hA + \mu_N Bg_I m_F \pm \frac{1}{2}(hAm_F + \mu_B Bg_J - \mu_N Bg_I))$$

$$\langle\pm|H|\mp\rangle = \frac{1}{2}hA\sqrt{(I+1/2)^2 - m_F^2}$$

Solving for the eigenvalues of this matrix, (as can be done by hand, or more easily, with a computer algebra system) we arrive at the energy shifts:

$$\Delta E_{F=I\pm 1/2} = -\frac{h\Delta W}{2(2I+1)} + \mu_N g_I m_F B \pm \frac{h\Delta W}{2}\sqrt{1+\frac{2m_F x}{I+1/2}+x^2}$$

$$x \equiv \frac{\mu_B B g_J - \mu_N B g_I}{h\Delta W} \qquad \Delta W = A\left(I+\frac{1}{2}\right)$$

where ΔW is the splitting (in units of Hz) between two hyperfine sublevels in the absence of magnetic field B,

x is referred to as the 'field strength parameter' (for $m=-(I+1/2)$ the square root is an exact square, and should be interpreted as $+(1-x)$)). This equation is known as the Breit-Rabi formula and is useful for systems with one valence electron in an s ($J=1/2$) level.

Note that index F in $\Delta E_{F=I\pm 1/2}$ should be considered not as total angular momentum of the atom but as asymptotic total angular momentum. It is equal to total angular momentum only if   0 otherwise eigenvectors corresponding different eigenvalues of the Hamiltonian are the superpositions of states with different F but equal (the only exceptions are $|F=I+1/2, m_F=\pm F\rangle$).

Applications

Astrophysics

George Ellery Hale was the first to notice the Zeeman effect in the solar spectra, indicating the existence of strong magnetic fields in sunspots. Such fields can be quite high, on the order of 0.1 tesla or higher. Today, the Zeeman effect is used to produce magnetograms showing the variation of magnetic field on the sun.

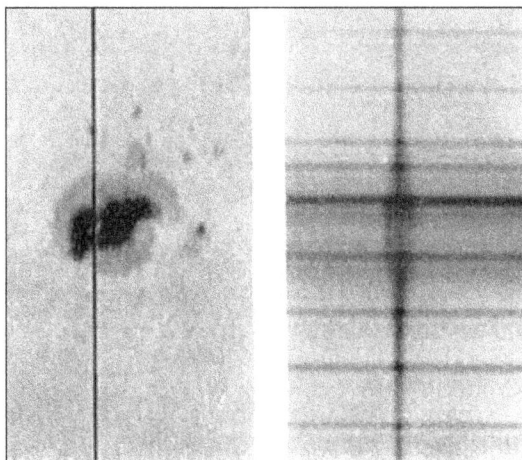
Zeeman effect on a sunspot spectral line.

Laser Cooling

The Zeeman effect is utilized in many laser cooling applications such as a magneto-optical trap and the Zeeman slower.

Zeeman-energy Mediated Coupling of Spin and Orbital Motions

Spin-orbit interaction in crystals is usually attributed to coupling of Pauli matrices σ to electron momentum k which exists even in the absence of magnetic field B. However, under the conditions of the Zeeman effect, when $\neq$, a similar interaction can be achieved by coupling σ to the electron coordinate r through the spatially inhomogeneous Zeeman Hamiltonian:

$$H_Z = \frac{1}{2}(\mathbf{B}\hat{g}\sigma),$$

where $\hat{g}$ is a tensorial Landé g-factor and either $B = B(r)$ or $\hat{g} = \hat{g}(\mathbf{r})$,, or both of them, depend on the electron coordinate r Such r -dependent Zeeman Hamiltonian $H_Z(\mathbf{r})$ couples electron spin σ to the operator r representing electron's orbital motion. Inhomogeneous field $\mathbf{B}(\mathbf{r})$ may be either a smooth field of external sources or fast-oscillating microscopic magnetic field in antiferromagnets. Spin-orbit coupling through macroscopically inhomogeneous field $\mathbf{B}(\mathbf{r})$ of nanomagnets is used for electrical operation of electron spins in quantum dots through electric dipole spin resonance, and driving spins by electric field due to inhomogeneous $\hat{g}$ (r)has been also demonstrated.

Hartree–Fock Method

The Hartree–Fock (HF) method is a method of approximation for the determination of the wave function and the energy of a quantum many-body system in a stationary state.

The Hartree–Fock method often assumes that the exact N-body wave function of the system can be approximated by a single Slater determinant (in the case where the particles are fermions) or by a single permanent (in the case of bosons) of N spin-orbitals. By invoking the variational method, one can derive a set of N-coupled equations for the N spin orbitals. A solution of these equations yields the Hartree–Fock wave function and energy of the system.

Especially in the older literature, the Hartree–Fock method is also called the self-consistent field method (SCF). In deriving what is now called the Hartree equation as an approximate solution of the Schrödinger equation, Hartreerequired the final field as computed from the charge distribution to be "self-consistent" with the assumed initial field. Thus, self-consistency was a requirement of the solution. The solutions to the non-linear Hartree–Fock equations also behave as if each particle is subjected to the mean field created by all other particles (the Fock operator below), and hence the terminology continued. The equations are almost universally solved by means of an iterative method, although the fixed-point iteration algorithm does not always converge. This solution scheme is not the only one possible and is not an essential feature of the Hartree–Fock method.

The Hartree–Fock method finds its typical application in the solution of the Schrödinger equation for atoms, molecules, nanostructures and solids but it has also found widespread use in nuclear physics. (See Hartree–Fock–Bogoliubov method for a discussion of its application in nuclear structure theory). In atomic structure theory, calculations may be for a spectrum with many excited energy levels and consequently the Hartree–Fock method for atoms assumes the wave function is a single configuration state function with well-defined quantum numbers and that the energy level is not necessarily the ground state.

For both atoms and molecules, the Hartree–Fock solution is the central starting point for most methods that describe the many-electron system more accurately.

Doubly occupied. Open-shell systems, where some of the electrons are not paired, can be dealt with by either the restricted open-shell or the unrestricted Hartree-Fock methods.

Early Semi-empirical Methods

The origin of the Hartree–Fock method dates back to the end of the 1920s, soon after the discovery of the Schrödinger equation in 1926. Douglas Hartree's methods were guided by some earlier, semi-empirical methods of the early 1920s (by E. Fues, R. B. Lindsay, and himself) set in the old quantum theory of Bohr.

In the Bohr model of the atom, the energy of a state with principal quantum number n is given in atomic units as $E = -1/n^2$. It was observed from atomic spectra that the energy levels of many-electron atoms are well described by applying a modified version of Bohr's formula. By introducing the quantum defect d as an empirical parameter, the energy levels of a generic atom were well approximated by the formula $E = -1/(n+d)^2$, in the sense that one could reproduce fairly well the observed transitions levels observed in the X-ray region (for example, see the empirical discussion and derivation in Moseley's law). The existence of a non-zero quantum defect was attributed to electron–electron repulsion, which clearly does not exist in the isolated hydrogen atom. This repulsion resulted in partial screening of the bare nuclear charge. These early researchers later introduced other potentials containing additional empirical parameters with the hope of better reproducing the experimental data.

Hartree Method

In 1927, D. R. Hartree introduced a procedure, which he called the self-consistent field method, to calculate approximate wave functions and energies for atoms and ions. Hartree sought to do away with empirical parameters and solve the many-body time-independent Schrödinger equation from fundamental physical principles, i.e., ab initio. His first proposed method of solution became known as the Hartree method, or Hartree product. However, many of Hartree's contemporaries did not understand the physical reasoning behind the Hartree method: it appeared to many people to contain empirical elements, and its connection to the solution of the many-body Schrödinger equation was unclear. However, in 1928 J. C. Slater and J. A. Gaunt independently showed that the Hartree method could be couched on a sounder theoretical basis by applying the variational principle to an ansatz (trial wave function) as a product of single-particle functions.

In 1930, Slater and V. A. Fock independently pointed out that the Hartree method did not respect the principle of antisymmetry of the wave function. The Hartree method used the Pauli exclusion principle in its older formulation, forbidding the presence of two electrons in the same quantum state. However, this was shown to be fundamentally incomplete in its neglect of quantum statistics.

Hartree-Fock

A solution to the lack of anti-symmetry in the Hartree method came when it was shown that a Slater determinant, a determinant of one-particle orbitals first used by Heisenberg and Dirac in 1926, trivially satisfies the antisymmetric property of the exact solution and hence is a suitable ansatz for applying the variational principle. The original Hartree method can then be viewed as an approximation to the Hartree–Fock method by neglecting exchange. Fock's original method relied heavily on group theory and was too abstract for contemporary physicists to understand and implement. In 1935, Hartree reformulated the method to be more suitable for the purposes of calculation.

The Hartree–Fock method, despite its physically more accurate picture, was little used until the advent of electronic computers in the 1950s due to the much greater computational demands over the early Hartree method and empirical models. Initially, both the Hartree method and the Hartree–Fock method were applied exclusively to atoms, where the spherical symmetry of the system allowed one to greatly simplify the problem. These approximate methods were (and are) often used together with the central field approximation, to impose the condition that electrons in the same shell have the same radial part, and to restrict the variational solution to be a spin eigenfunction. Even so, calculating a solution by hand using the Hartree–Fock equations for a medium-sized atom was laborious; small molecules required computational resources far beyond what was available before 1950.

Hartree–Fock Algorithm

The Hartree–Fock method is typically used to solve the time-independent Schrödinger equation for a multi-electron atom or molecule as described in the Born–Oppenheimer approximation. Since there are no known analytic solutions for many-electron systems (there are solutions for one-electron systems such as hydrogenic atoms and the diatomic hydrogen cation), the problem is solved numerically. Due to the nonlinearities introduced by the Hartree–Fock approximation, the equations are solved using a nonlinear method such as iteration, which gives rise to the name "self-consistent field method".

Approximations

The Hartree–Fock method makes five major simplifications in order to deal with this task:

- The Born–Oppenheimer approximation is inherently assumed. The full molecular wave function is actually a function of the coordinates of each of the nuclei, in addition to those of the electrons.

- Typically, relativistic effects are completely neglected. The momentum operator is assumed to be completely non-relativistic.

- The variational solution is assumed to be a linear combination of a finite number of basis functions, which are usually (but not always) chosen to be orthogonal. The finite basis set is assumed to be approximately complete.

- Each energy eigenfunction is assumed to be describable by a single Slater determinant, an antisymmetrized product of one-electron wave functions (i.e., orbitals).

- The mean-field approximation is implied. Effects arising from deviations from this assumption are neglected. These effects are often collectively used as a definition of the term electron correlation. However, the label "electron correlation" strictly spoken encompasses both Coulomb correlation and Fermi correlation, and the latter is an effect of electron exchange, which is fully accounted for in the Hartree–Fock method. Stated in this terminology, the method only neglects the Coulomb correlation. However, this is an important flaw, accounting for (among others) Hartree–Fock's inability to capture London dispersion.

Relaxation of the last two approximations give rise to many so-called post-Hartree–Fock methods.

Variational Optimization of Orbitals

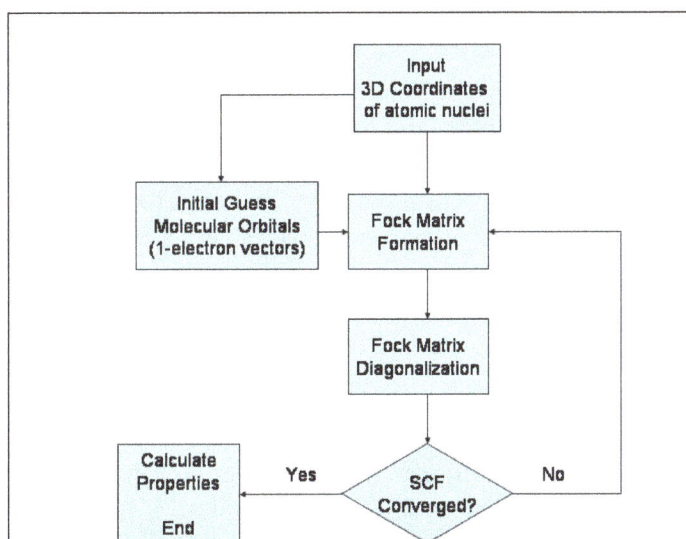

Algorithmic flowchart illustrating the Hartree–Fock method.

The variational theorem states that for a time-independent Hamiltonian operator, any trial wave function will have an energy expectation value that is greater than or equal to the true ground-state wave function corresponding to the given Hamiltonian. Because of this, the Hartree–Fock energy is an upper bound to the true ground-state energy of a given molecule. In the context of the Hartree–Fock method, the best possible solution is at the Hartree–Fock limit; i.e., the limit of the Hartree–Fock energy as the basis set approaches completeness. (The other is the full-CI limit, where the last two approximations of the Hartree–Fock theory as described above are completely undone. It is only when both limits are attained that the exact solution, up to the Born–Oppenheimer approximation, is obtained). The Hartree–Fock energy is the minimal energy for a single Slater determinant.

The starting point for the Hartree–Fock method is a set of approximate one-electron wave functions known as spin-orbitals. For an atomic orbital calculation, these are typically the orbitals for a hydrogen-like atom (an atom with only one electron, but the appropriate nuclear charge). For a molecular orbital or crystalline calculation, the initial approximate one-electron wave functions are typically a linear combination of atomic orbitals (LCAO).

The orbitals above only account for the presence of other electrons in an average manner. In the Hartree–Fock method, the effect of other electrons are accounted for in a mean-field theory context. The orbitals are optimized by requiring them to minimize the energy of the respective

Slater determinant. The resultant variational conditions on the orbitals lead to a new one-electron operator, the Fock operator. At the minimum, the occupied orbitals are eigensolutions to the Fock operator via a unitary transformation between themselves. The Fock operator is an effective one-electron Hamiltonian operator being the sum of two terms. The first is a sum of kinetic-energy operators for each electron, the internuclear repulsion energy, and a sum of nuclear–electronic Coulombic attraction terms. The second are Coulombic repulsion terms between electrons in a mean-field theory description; a net repulsion energy for each electron in the system, which is calculated by treating all of the other electrons within the molecule as a smooth distribution of negative charge. This is the major simplification inherent in the Hartree–Fock method.

Since the Fock operator depends on the orbitals used to construct the corresponding Fock matrix, the eigenfunctions of the Fock operator are in turn new orbitals, which can be used to construct a new Fock operator. In this way, the Hartree–Fock orbitals are optimized iteratively until the change in total electronic energy falls below a predefined threshold. In this way, a set of self-consistent one-electron orbitals is calculated. The Hartree–Fock electronic wave function is then the Slater determinant constructed from these orbitals. Following the basic postulates of quantum mechanics, the Hartree–Fock wave function can then be used to compute any desired chemical or physical property within the framework of the Hartree–Fock method and the approximations employed.

Mathematical Formulation

The Fock Operator

Because the electron–electron repulsion term of the molecular Hamiltonian involves the coordinates of two different electrons, it is necessary to reformulate it in an approximate way. Under this approximation (outlined under Hartree–Fock algorithm), all of the terms of the exact Hamiltonian except the nuclear–nuclear repulsion term are re-expressed as the sum of one-electron operators outlined below, for closed-shell atoms or molecules (with two electrons in each spatial orbital). The (1) following each operator symbol simply indicates that the operator is 1-electron in nature.

$$\hat{F}[\{\phi_j\}](1) = \hat{H}^{core}(1) + \sum_{j=1}^{N/2}[2\hat{J}_j(1) - \hat{K}_j(1)],$$

where,

$$\hat{F}[\{\phi_j\}](1)$$

is the one-electron Fock operator generated by the orbitals ϕ_j,

$$\hat{H}^{core}(1) = -\frac{1}{2}\nabla_1^2 - \sum_\alpha \frac{Z_\alpha}{r_{1\alpha}},$$

is the one-electron core Hamiltonian,

$$\hat{J}_j(1),$$

is the Coulomb operator, defining the electron–electron repulsion energy due to each of the two electrons in the j-th orbital,

$$\hat{K}_j(1)_,$$

is the exchange operator, defining the electron exchange energy due to the antisymmetry of the total N-electron wave function. This "exchange energy" operator $\hat{K}$ is simply an artifact of the Slater determinant. Finding the Hartree–Fock one-electron wave functions is now equivalent to solving the eigenfunction equation:

$$\hat{F}(1)\phi_i(1) = \epsilon_i\phi_i(1),$$

where $\phi_i(1)$ are a set of one-electron wave functions, called the Hartree–Fock molecular orbitals.

Linear Combination of Atomic Orbitals

Typically, in modern Hartree–Fock calculations, the one-electron wave functions are approximated by a linear combination of atomic orbitals. These atomic orbitals are called Slater-type orbitals. Furthermore, it is very common for the "atomic orbitals" in use to actually be composed of a linear combination of one or more Gaussian-type orbitals, rather than Slater-type orbitals, in the interests of saving large amounts of computation time.

Various basis sets are used in practice, most of which are composed of Gaussian functions. In some applications, an orthogonalization method such as the Gram–Schmidt process is performed in order to produce a set of orthogonal basis functions. This can in principle save computational time when the computer is solving the Roothaan–Hall equations by converting the overlap matrix effectively to an identity matrix. However, in most modern computer programs for molecular Hartree–Fock calculations this procedure is not followed due to the high numerical cost of orthogonalization and the advent of more efficient, often sparse, algorithms for solving the generalized eigenvalue problem, of which the Roothaan–Hall equations are an example.

Numerical Stability

Numerical stability can be a problem with this procedure and there are various ways of combating this instability. One of the most basic and generally applicable is called F-mixing or damping. With F-mixing, once a single-electron wave function is calculated, it is not used directly. Instead, some combination of that calculated wave function and the previous wave functions for that electron is used, the most common being a simple linear combination of the calculated and immediately preceding wave function. A clever dodge, employed by Hartree, for atomic calculations was to increase the nuclear charge, thus pulling all the electrons closer together. As the system stabilised, this was gradually reduced to the correct charge. In molecular calculations a similar approach is sometimes used by first calculating the wave function for a positive ion and then to use these orbitals as the starting point for the neutral molecule. Modern molecular Hartree–Fock computer programs use a variety of methods to ensure convergence of the Roothaan–Hall equations.

Weaknesses, Extensions and Alternatives

Of the five simplifications outlined in the section "Hartree–Fock algorithm", the fifth is typically the most important. Neglect of electron correlation can lead to large deviations from experimental results. A number of approaches to this weakness, collectively called post-Hartree–Fock methods, have been devised to include electron correlation to the multi-electron wave function. One of these approaches, Møller–Plesset perturbation theory, treats correlation as a perturbation of the Fock operator. Others expand the true multi-electron wave function in terms of a linear combination of Slater determinants-such as multi-configurational self-consistent field, configuration interaction, quadratic configuration interaction, and complete active space SCF (CASSCF). Still others (such as variational quantum Monte Carlo) modify the Hartree–Fock wave function by multiplying it by a correlation function ("Jastrow" factor), a term which is explicitly a function of multiple electrons that cannot be decomposed into independent single-particle functions.

An alternative to Hartree–Fock calculations used in some cases is density functional theory, which treats both exchange and correlation energies, albeit approximately. Indeed, it is common to use calculations that are a hybrid of the two methods-the popular B3LYP scheme is one such hybrid functional method. Another option is to use modern valence bond methods.

Brillouin's Theorem

In quantum chemistry, Brillouin's theorem, proposed by the French physicist Léon Brillouin in 1934, states that given a self-consistent optimized Hartree-Fock wavefunction $|\psi_0\rangle$, the matrix element of the Hamiltonian between the ground state and a single excited determinant (i.e. one where an occupied orbital a is replaced by a virtual orbital r) must be zero.

$$\langle \psi_0 | \hat{H} | \psi_a^r \rangle = 0$$

This theorem is important in constructing a configuration interaction method, among other applications.

Another interpretation of the theorem is that the ground electronic states solved by one-particle methods (such as HF or DFT) already imply configuration interaction of the ground-state configuration with the singly excited ones. That renders their further inclusion into the CI expansion redundant.

Proof:

The electronic Hamiltonian of the system can be divided into two parts: one consisting of one-electron operators $h(1) = -\frac{1}{2}\nabla_1^2 - \sum_\alpha \frac{Z_\alpha}{r_{1\alpha}}$ and the other of two-electron operators $\sum_j |r_i - r_j|^{-1}$.. In methods of wavefunction-based quantum chemistry which include the electron correlation into the model, the wavefunction is expressed as a sum of series consisting of different Slater determinants (i.e., a linear combination of such determinants). In the simplest case of configuration interaction (as well as in other single-reference multielectron-basis set methods, like MPn, etc)., all the determinants contain the same one-electron functions, or orbitals, and differ just by occupation of these orbitals by electrons. The source of these orbitals is the converged Hartree–Fock calculation, which gives the so-called reference determinant $|\psi_0\rangle$ with all the electrons occupying energetically lowest states among the available. All other determinants are then made by formally "exciting" the reference one (one or more electrons are cleared from one-electron states occupied in $|\psi_0\rangle$ and put into states unoccupied in $|\psi_0\rangle$. As the orbitals remain the same, we can simply transition from the many-electron state basis $(|\psi_0\rangle|\psi_a^r\rangle,|\psi_{ab}^{rs}\rangle)$ to the one-electron state basis (which was used for Hartree–Fock: $|a\rangle,|b\rangle,|r\rangle,|s\rangle,\ldots$), greatly improving the efficiency of calculations. For this transition, we apply the Slater–Condon rules and evaluate:

$$\left\langle \psi_0 \mid \hat{H} \mid \psi_a^r \right\rangle = \langle a \mid h \mid r \rangle + \sum_b \langle ab \| rb \rangle = \langle a \mid h \mid r \rangle + \sum_b \left(\langle ab \mid rb \rangle - \langle ab \mid br \rangle \right) = \langle a \mid h \mid r \rangle + \sum_b \left(\langle a \mid 2\hat{J}_b - \hat{K}_b \mid r \rangle \right)$$

which we recognize is simply an off-diagonal element of the Fock matrix $\langle \chi_a \mid \hat{F} \mid \chi_r \rangle$. But the reference wave function was obtained by the Hartree–Fock calculation, or the SCF procedure, whole point of which was to diagonalize the Fock matrix. Hence for an optimized wavefunction this off-diagonal element must be zero.

This can be made evident also if we multiply both sides of a Hartree–Fock equation:

$$\hat{F}\chi_r = \epsilon_r \chi_r,$$

by $\chi_a^*(\vec{r})$ and integrate over the electronic coordinate:

$$\int_{-\infty}^{\infty} \chi_a^*(\vec{r})\hat{F}\chi_r(\vec{r})d^3\vec{r} = \epsilon_r \int_{-\infty}^{\infty} \chi_a^*(\vec{r})\chi_r(\vec{r})d^3\vec{r}.$$

As the Fock matrix has already been diagonalized, the states $\chi_a(\vec{r})$ and $\chi_a(\vec{r})$ are the eigenstates of the Fock operator, and as such are orthogonal; thus their overlap is zero. It makes all the right-hand side of the equation zero:

$$\int_{-\infty}^{\infty} \chi_a^*(\vec{r})\hat{F}\chi_r(\vec{r})d^3\vec{r} = \langle \psi_0 \mid \hat{H} \mid \psi_a^r \rangle = 0,$$

which proves the Brillouin's theorem.

The theorem have also been proven directly from the variational principle (by Mayer) and is essentially equivalent to the Hartree–Fock equations in general.

Koopman's Theorem

Koopman's theorem states that in closed-shell Hartree–Fock theory (HF), the first ionization energy of a molecular system is equal to the negative of the orbital energy of the highest occupied molecular orbital (HOMO). This theorem is named after Tjalling Koopmans, who published this result in 1934.

Koopman's theorem is exact in the context of restricted Hartree–Fock theory if it is assumed that the orbitals of the ion are identical to those of the neutral molecule (the frozen orbital approximation).

Ionization energies calculated this way are in qualitative agreement with experiment – the first ionization energy of small molecules is often calculated with an error of less than two electron volts. Therefore, the validity of Koopman's theorem is intimately tied to the accuracy of the underlying Hartree–Fock wavefunction. The two main sources of error are orbital relaxation, which refers to the changes in the Fock operator and Hartree–Fock orbitals when changing the number of electrons in the system, and electron correlation, referring to the validity of representing the entire many-body wavefunction using the Hartree–Fock wavefunction, i.e. a single Slater determinant composed of orbitals that are the eigenfunctions of the corresponding self-consistent Fock operator.

Empirical comparisons with experimental values and higher-quality ab initio calculations suggest that in many cases, but not all, the energetic corrections due to relaxation effects nearly cancel the corrections due to electron correlation.

A similar theorem exists in density functional theory (DFT) for relating the exact first vertical ionization energy and electron affinity to the HOMO and LUMO energies, although both the derivation and the precise statement differ from that of Koopman's theorem. Ionization energies calculated from DFT orbital energies are usually poorer than those of Koopman's theorem, with errors much larger than two electron volts possible depending on the exchange-correlation approximation employed. The LUMO energy shows little correlation with the electron affinity with typical approximations. The error in the DFT counterpart of Koopman's theorem is a result of the approximation employed for the exchange correlation energy functional so that, unlike in HF theory, there is the possibility of improved results with the development of better approximations.

Generalizations

While Koopman's theorem was originally stated for calculating ionization energies from restricted (closed-shell) Hartree–Fock wavefunctions, the term has since taken on a more generalized meaning as a way of using orbital energies to calculate energy changes due to changes in the number of electrons in a system.

Ground-state and Excited-state Ions

Koopman's theorem applies to the removal of an electron from any occupied molecular orbital to form a positive ion. Removal of the electron from different occupied molecular orbitals leads to the ion in different electronic states. The lowest of these states is the ground state and this often, but not always, arises from removal of the electron from the HOMO. The other states are excited electronic states.

For example, the electronic configuration of the H_2O molecule is $(1a_1)^2 (2a_1)^2 (1b_2)^2 (3a_1)^2 (1b_1)^2$, where the symbols a_1, b_2 and b_1 are orbital labels based on molecular symmetry. From Koopman's theorem the energy of the $1b_1$ HOMO corresponds to the ionization energy to form the H_2O^+ ion in its ground state $(1a_1)^2 (2a_1)^2 (1b_2)^2 (3a_1)^2 (1b_1)^1$. The energy of the second-highest MO $3a_1$ refers to the ion in the excited state $(1a_1)^2 (2a_1)^2 (1b_2)^2 (3a_1)^1 (1b_1)^2$, and so on. In this case the order of the ion electronic states corresponds to the order of the orbital energies. Excited-state ionization energies can be measured by photoelectron spectroscopy.

For H_2O, the near-Hartree–Fock orbital energies (with sign changed) of these orbitals are $1a_1$ 559.5, $2a_1$ 36.7 $1b_2$ 19.5, $3a_1$ 15.9 and $1b_1$ 13.8 eV. The corresponding ionization energies are 539.7, 32.2, 18.5, 14.7 and 12.6 eV. As explained above, the deviations are due to the effects of orbital relaxation as well as differences in electron correlation energy between the molecular and the various ionized states.

For N_2 in contrast, the order of orbital energies is not identical to the order of ionization energies. Near-Hartree-Fock calculations with a large basis set indicate that the $1\pi_u$ bonding orbital is the HOMO. However the lowest ionization energy corresponds to removal of an electron from the $3\sigma_g$ bonding orbital. In this case the deviation is attributed primarily to the difference in correlation energy between the two orbitals.

For Electron Affinities

It is sometimes claimed that Koopman's theorem also allows the calculation of electron affinities as the energy of the lowest unoccupied molecular orbitals (LUMO) of the respective systems. However, Koopman's original paper makes no claim with regard to the significance of eigenvalues of the Fock operator other than that corresponding to the HOMO. Nevertheless, it is straightforward to generalize the original statement of Koopman's to calculate the electron affinity in this sense.

Calculations of electron affinities using this statement of Koopman's theorem have been criticized on the grounds that virtual (unoccupied) orbitals do not have well-founded physical interpretations, and that their orbital energies are very sensitive to the choice of basis set used in the calculation. As the basis set becomes more complete; more and more "molecular" orbitals that are not really on the molecule of interest will appear, and care must be taken not to use these orbitals for estimating electron affinities.

Comparisons with experiment and higher-quality calculations show that electron affinities predicted in this manner are generally quite poor.

For Open-shell Systems

Koopman's theorem is also applicable to open-shell systems. It was previously believed that this was only in the case for removing the unpaired electron, but the validity of Koopman's theorem for ROHF in general has been proven provided that the correct orbital energies are used. The spin up (alpha) and spin down (beta) orbital energies do not necessarily have to be the same.

Counterpart in Density Functional Theory

Kohn–Sham (KS) density functional theory (KS-DFT) admits its own version of Koopman's theorem (sometimes called the DFT-Koopman's theorem) very similar in spirit to that of Hartree-Fock theory. The theorem equates the first (vertical) ionization energy I of a system of N electrons to the negative of the corresponding KS HOMO energy ϵ_H. More generally, this relation is true even when the KS systems describes a zero-temperature ensemble with non-integer number of electrons $IN + \delta N$ for integer N and $\delta N \to 0$. When considering $IN + \delta N$ electrons the infinitesimal excess charge enters the KS LUMO of the N electron system but then the exact KS potential jumps by a constant known as the "derivative discontinuity". It can be argued that the vertical electron affinity is equal exactly to the negative of the sum of the LUMO energy and the derivative discontinuity.

Unlike the approximate status of Koopman's theorem in Hartree Fock theory (because of the neglect of orbital relaxation), in the exact KS mapping the theorem is exact, including the effect of orbital relaxation. A sketchy proof of this exact relation goes in three stages. First, for any finite system I determines the $|\mathbf{r}| \to \infty$ asymptotic form of the density, which decays as $n(\mathbf{r}) \to \exp\left(-2\sqrt{\frac{2m_e}{\hbar}} I \, |\mathbf{r}|\right)$. Next, as a corollary (since the physically interacting system has the same density as the KS system), both must have the same ionization energy. Finally, since the KS potential is zero at infinity, the ionization energy of the KS system is, by definition, the negative of its HOMO energy and thus finally: $\epsilon_H = -I$.

While these are exact statements in the formalism of DFT, the use of approximate exchange-correlation potentials makes the calculated energies approximate and often the orbital energies are very different from the corresponding ionization energies (even by several eV!).

A tuning procedure is able to "impose" Koopman's theorem on DFT approximations thereby improving many of its related predictions in actual applications.

In approximate DFTs one can estimate to high degree of accuracy the deviance from Koopman's theorem using the concept of energy curvature.

It provides excitation energies to zeroth-order and $\dfrac{\partial E}{\partial n_i} = \varepsilon_i$.

Hydrogen Atom

The hydrogen atom is the simplest atom in nature and, therefore, a good starting point to study atoms and atomic structure. The hydrogen atom consists of a single negatively charged electron that moves about a positively charged proton. In Bohr's model, the electron is pulled around the proton in a perfectly circular orbit by an attractive Coulomb force. The proton is approximately 1800 times more massive than the electron, so the proton moves very little in response to the force on the proton by the electron. (This is analogous to the Earth-Sun system, where the Sun moves very little in response to the force exerted on it by Earth). An explanation of this effect using Newton's laws is given in Photons and Matter Waves.

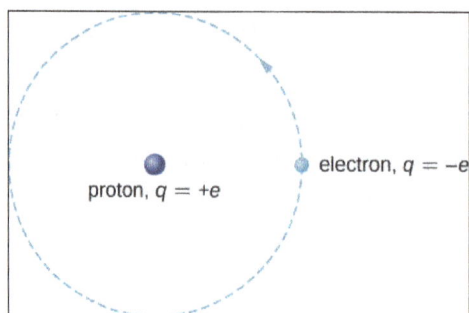

A representation of the Bohr model of the hydrogen atom.

With the assumption of a fixed proton, we focus on the motion of the electron.

In the electric field of the proton, the potential energy of the electron is:

$$U(r) = -k\frac{e^2}{r},$$

where $k = 1/4\pi\varepsilon_0$ and r is the distance between the electron and the proton. the force on an object is equal to the negative of the gradient (or slope) of the potential energy function. For the special case of a hydrogen atom, the force between the electron and proton is an attractive Coulomb force.

Notice that the potential energy function $U(r)$ does not vary in time. As a result, Schrödinger's equation of the hydrogen atom reduces to two simpler equations: one that depends only on space (x, y, z) and another that depends only on time (t) the space-dependent equation:

$$\frac{-\hbar}{2m_e}\left(\frac{\partial^2\psi}{\partial x^2} + \frac{\partial^2\psi}{\partial y^2} + \frac{\partial^2\psi}{\partial z^2}\right) - k\frac{e^2}{r}\psi = E\psi,$$

where $\psi = psi(x, y, z)$ is the three-dimensional wave function of the electron, meme is the mass of the electron, and E is the total energy of the electron. Recall that the total wave function $\varnothing(x, y, z, t)$, is the product of the space-dependent wave function $\psi = \psi(x, y, z)$ and the time-dependent wave function $\varphi = \varphi(t)$.

In addition to being time-independent, $U(r)$ is also spherically symmetrical. This suggests that we may solve Schrödinger's equation more easily if we express it in terms of the spherical coordinates (r, θ, ϕ) instead of rectangular coordinates (x,y,z). A spherical coordinate system is shown in figure. In spherical coordinates, the variable r is the radial coordinate, θ is the polar angle (relative to the vertical z-axis), and φ is the azimuthal angle (relative to the x-axis). The relationship between spherical and rectangular coordinates is $x = r\sin\theta\cos\phi$, $y = r\sin\theta\sin\phi$, $z = r\cos\theta$.

The factor $r\sin\theta$ is the magnitude of a vector formed by the projection of the polar vector onto the xy-plane. Also, the coordinates of x and y are obtained by projecting this vector onto the x- and y-axes, respectively. The inverse transformation gives:

$$r = \sqrt{x^2 + y^2 + z^2}, \theta = \cos^{-1}\left(\frac{z}{r}\right), \phi = \cos^{-1}\left(\frac{x}{\sqrt{x^2 + y^2}}\right)$$

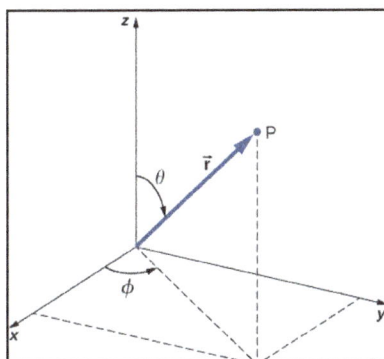

The relationship between the spherical and rectangular coordinate systems.

However, due to the spherical symmetry of $U(r)$, this equation reduces to three simpler equations: one for each of the three coordinates (r, θ, and φ). Solutions to the time-independent wave function are written as a product of three functions:

$$\psi(r,\theta,\phi) = R(r)\Theta(\theta)\Phi(\phi),$$

where R is the radial function dependent on the radial coordinate r only; Θ is the polar function dependent on the polar coordinate θ only; and Φ is the phi function of φ only. Valid solutions to Schrödinger's equation $\psi(r,\theta,\phi)$ are labeled by the quantum numbers n, l, and m.

n: principal quantum number

l: angular momentum quantum number

m: angular momentum projection quantum number

The radial function R depends only on n and l; the polar function Θ depends only on l and m; and the phi function Φ depends only on m. The dependence of each function on quantum numbers is indicated with subscripts:

$$\psi_{nlm}(r,\theta,\phi) = R_{nl}(r)\Theta_{lm}(\theta)\Phi_m(\phi)$$

Not all sets of quantum numbers (n, l, m) are possible. For example, the orbital angular quantum number l can never be greater or equal to the principal quantum number $n (l < n)$ .Specifically, we have:

$$n = 1, 2, 3, ...$$
$$l = 0, 1, 2, ..., (n-1)$$
$$m = -l, (-l+1), ..., 0, ..., (+l-1), +l$$

Notice that for the ground state, $n=1$ $l=0$ and m=0. In other words, there is only one quantum state with the wave function for $n=1$, and it is ψ_{100}. However, for n = 2, we have:

$$l = 0, m = 0$$
$$l = 1, m = -1, 0, 1.$$

Therefore, the allowed states for the n=2 state are $\psi_{200}, \psi_{21-1}, \psi_{210},$ and ψ_{211}. Example wave functions for the hydrogen atom are given in table. Note that some of these expressions contain the letter i, which represents $\sqrt{-1}$. When probabilities are calculated, these complex numbers do not appear in the final answer.

Wave Functions of the Hydrogen Atom	
$n = 1, l = 0, m_l = 0$	$\psi_{100} = \dfrac{1}{\sqrt{\pi}} \dfrac{1}{a_0^{3/2}} e^{-r/a_0}$
$n = 2, l = 0, m = 0$	$\psi_{200} = \dfrac{1}{4\sqrt{2\pi}} \dfrac{1}{a_0^{3/2}} (2 - \dfrac{r}{a_0}) e^{-r/2a_0}$

$n = 2, l = 1, m_l = -1$	$\psi_{21-1} = \dfrac{1}{8\sqrt{\pi}} \dfrac{1}{a_0^{3/2}} \dfrac{r}{a_0} e^{-r/2a_0} \sin\theta\, e^{-i\phi}$
$n = 2, l = 1, m_l = 0$	$\psi_{210} = \dfrac{1}{4\sqrt{2\pi}} \dfrac{1}{a_0^{3/2}} \dfrac{r}{a_0} e^{-r/2a_0} \cos\theta$
$n = 2, l = 1, m_l = 1$	$\psi_{211} = \dfrac{1}{8\sqrt{\pi}} \dfrac{1}{a_0^{3/2}} \dfrac{r}{a_0} e^{-r/2a_0} \sin\theta\, e^{i\phi}$

Physical Significance of the Quantum Numbers

Each of the three quantum numbers of the hydrogen atom (n, l, m) is associated with a different physical quantity. The principal quantum number n is associated with the total energy of the electron, E_n. According to Schrödinger's equation:

$$E_n = -\left(\frac{m_e k^2 e^4}{2^2}\right)\left(\frac{1}{n^2}\right) = -E_0\left(\frac{1}{n^2}\right),$$

where $E_0 = -13.6eV$. Notice that this expression is identical to that of Bohr's model. As in the Bohr model, the electron in a particular state of energy does not radiate.

The angular momentum orbital quantum number l is associated with the orbital angular momentum of the electron in a hydrogen atom. Quantum theory tells us that when the hydrogen atom is in the state ψ_{nlm}, the magnitude of its orbital angular momentum is:

$$L = \sqrt{l(l+1)}\hbar,$$

Where,

$$l = 0, 1, 2, ..., (n-1).$$

This result is slightly different from that found with Bohr's theory, which quantizes angular momentum according to the rule $L = n$, where n=1,2,3.

Quantum states with different values of orbital angular momentum are distinguished using spectroscopic notation. The designations s, p, d, and f result from early historical attempts to classify atomic spectral lines. (The letters stand for sharp, principal, diffuse, and fundamental, respectively). After f, the letters continue alphabetically.

The ground state of hydrogen is designated as the 1s state, where "1" indicates the energy level (n=1) and "s" indicates the orbital angular momentum state $(l = 0)$. When n=2,l can be either 0 or 1. The n=2, l=0 state is designated "2s". The n=2, l=1 state is designated "2p". When n=3 l can be 0, 1, or 2, and the states are 3s, 3p, and 3d, respectively. Notation for other quantum states is given in table.

The angular momentum projection quantum number m is associated with the azimuthal angle ϕ and is related to the z-component of orbital angular momentum of an electron in a hydrogen atom. This component is given by:

$L_z = m\hbar$,

Where,

$m = -l, -l+1, ..., 0, ..., + l-1, l.$

The z-component of angular momentum is related to the magnitude of angular momentum by:

$L_z = L\cos\theta,$

where θ is the angle between the angular momentum vector and the z-axis. Note that the direction of the z-axis is determined by experiment - that is, along any direction, the experimenter decides to measure the angular momentum. For example, the z-direction might correspond to the direction of an external magnetic field. The relationship between L_z and L is given in figure.

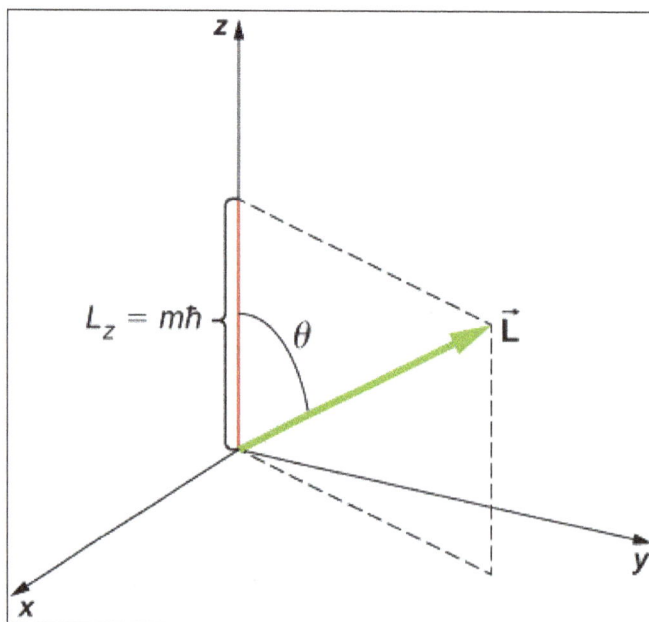

The z-component of angular momentum is quantized with its own quantum number m.

Spectroscopic Notation and Orbital Angular Momentum			
Orbital Quantum Number l	Angular Momentum	State	Spectroscopic Name
0	0	s	Sharp
1	$\sqrt{2}\hbar$	p	Principal
2	$\sqrt{6}\hbar$	d	Diffuse

Spectroscopic Notation and Orbital Angular Momentum			
Orbital Quantum Number l	Angular Momentum	State	Spectroscopic Name
3	$\sqrt{12}\hbar$	f	Fundamental
4	$\sqrt{20}\hbar$	g	
5	$\sqrt{30}\hbar$	h	

Spectroscopic Description of Quantum States						
	$l=0$	$l=1$	$l=2$	$l=3$	$l=4$	$l=5$
n=1	1s					
n=2	2s	2p				
n=3	3s	3p	3d			
n=4	4s	4p	4d	4f		
n=5	5s	5p	5d	5f	5g	
n=6	6s	6p	6d	6f	6g	6h

The quantization of L_z is equivalent to the quantization of |theta. Substituting $\sqrt{l(l+1)}\hbar$ for L and m for L_z into this equation, we find:

$$m\hbar = \sqrt{l(l+1)}\hbar cos\theta.$$

Thus, the angle θ is quantized with the particular values:

$$\theta = cos^{-1}\left(\frac{m}{\sqrt{l(l+1)}}\right)$$

Notice that both the polar angle (θ θ) and the projection of the angular momentum vector onto an arbitrary z-axis (L$_z$) are quantized.

The quantization of the polar angle for the *l=3* state is shown in figure. The orbital angular momentum vector lies somewhere on the surface of a cone with an opening angle θ relative to the z-axis (unless m=0 in which case θ = 90° and the vector points are perpendicular to the z-axis).

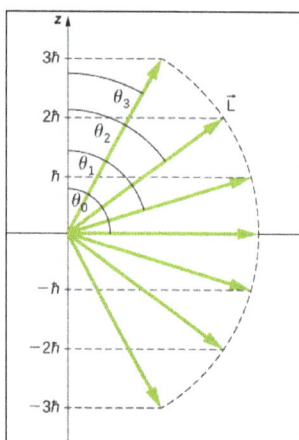

The quantization of orbital angular momentum. Each vector lies on the surface of a cone with axis along the z-axis.

A detailed study of angular momentum reveals that we cannot know all three components simultaneously. In the previous section, the z-component of orbital angular momentum has definite values that depend on the quantum number m. This implies that we cannot know both x- and y-components of angular momentum, L_x and L_y, with certainty. As a result, the precise direction of the orbital angular momentum vector is unknown.

Check Your Understanding Can the magnitude of LzLz ever be equal to L?

No. The quantum number $m = -l, -l + 1, ..., 0, ..., l-1, l$. Thus, the magnitude of Lz is always less than L because $< \sqrt{l(l+1)}$.

Using the Wave Function to make Predictions

we can use quantum mechanics to make predictions about physical events by the use of probability statements. It is therefore proper to state, "An electron is located within this volume with this probability at this time," but not, "An electron is located at the position (x, y, z) at this time". To determine the probability of finding an electron in a hydrogen atom in a particular region of space, it is necessary to integrate the probability density |ψnlm|2|ψnlm|2 over that region:

$$Probability = \int_{volume} | \psi_{nlm} |^2 \, dV,$$

where dV is an infinitesimal volume element. If this integral is computed for all space, the result is 1, because the probability of the particle to be located somewhere is 100% (the normalization condition). In a more advanced course on modern physics, you will find that $| \psi_{nlm} |^2 = \psi_{nlm}^* \psi_{nlm}$, where ψ_{nlm}^* is the complex conjugate. This eliminates the occurrences $i = \sqrt{-1}$ in the above calculation.

Consider an electron in a state of zero angular momentum ($l=0$). In this case, the electron's wave function depends only on the radial coordinate r. (Refer to the states $\psi_{100} \psi_{200}$ in Table) The infinitesimal volume element corresponds to a spherical shell of radius r and infinitesimal thickness dr, written as:

$$dV = 4\pi r^2 dr.$$

The probability of finding the electron in the region r to r +d r ("at approximately r") is,

$$P(r)dr = |\psi_{n00}|^2 \, 4\pi r^2 dr.$$

Here $P(r)$ is called the radial probability density function (a probability per unit length). For an electron in the ground state of hydrogen, the probability of finding an electron in the region r to $r + dr$ is:

$$|\psi_{n00}|^2 \, 4\pi r^2 dr = (4/a_)^3) \, r^2 exp(-2r/a_0) dr,$$

where $a_0 = 0.5$ angstroms. The radial probability density function $P(r)$ is plotted in figure. The area under the curve between any two radial positions, say r_1 and r_2, gives the probability of finding the electron in that radial range. To find the most probable radial position, we set the first derivative of this function to zero $(dP/dr=0)$ and solve for r. The most probable radial position is not equal to the average or expected value of the radial position because $|\psi_{n00}|^2$ is not symmetrical about its peak value.

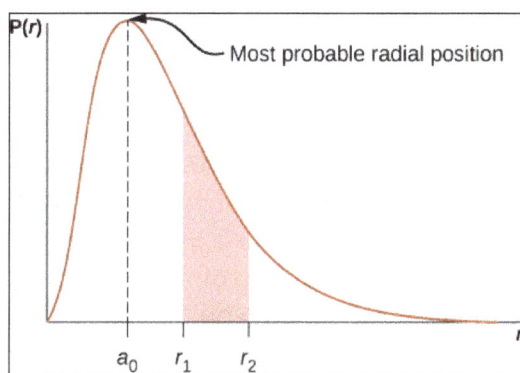

The radial probability density function for the ground state of hydrogen.

If the electron has orbital angular momentum $(l \neq 0)$, then the wave functions representing the electron depend on the angles θ and ϕ; that is, $\psi_{nlm} = \psi_{nlm}(r, \theta, \phi)$. Atomic orbitals for three states with n=2 and l=1are shown in figure. An atomic orbital is a region in space that encloses a certain percentage (usually 90%) of the electron probability. (Sometimes atomic orbitals are referred to as "clouds" of probability). Notice that these distributions are pronounced in certain directions. This directionality is important to chemists when they analyze how atoms are bound together to form molecules.

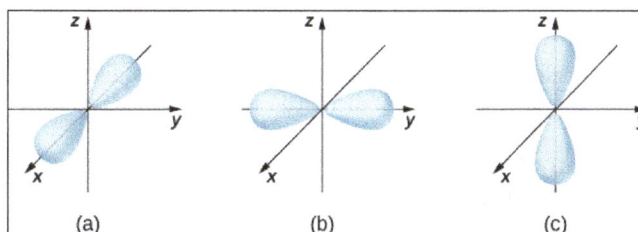

The probability density distributions for three states with n=2n=2 and l=1l=1.
The distributions are directed along the (a) x-axis, (b) y-axis, and (c) z-axis.

A slightly different representation of the wave function is given in figure. In this case, light and dark regions indicate locations of relatively high and low probability, respectively. In contrast to the Bohr model of the hydrogen atom, the electron does not move around the proton nucleus in a well-defined path. Indeed, the uncertainty principle makes it impossible to know how the electron gets from one place to another.

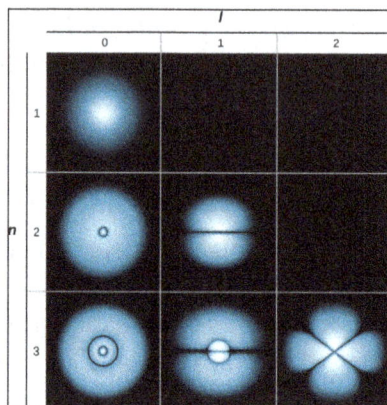

Probability clouds for the electron in the ground state and several excited states of hydrogen. The probability of finding the electron is indicated by the shade of color; the lighter the coloring, the greater the chance of finding the electron.

Schrödinger Equation for Hydrogen Atom

The hydrogen atom, consisting of an electron and a proton, is a two-particle system, and the internal motion of two particles around their center of mass is equivalent to the motion of a single particle with a reduced mass. This reduced particle is located at r, where r is the vector specifying the position of the electron relative to the position of the proton. The length of r is the distance between the proton and the electron, and the direction of r and the direction of r is given by the orientation of the vector pointing from the proton to the electron. Since the proton is much more massive than the electron, we will assume that the reduced mass equals the electron mass and the proton is located at the center of mass.

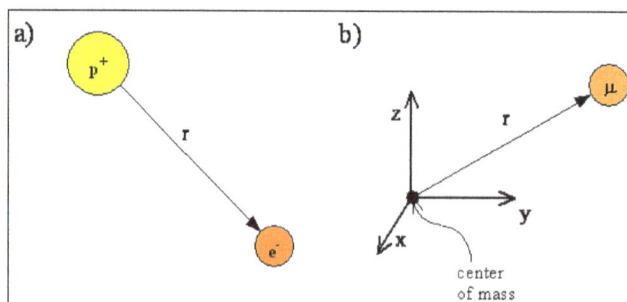

The proton (p+) and electron (e–) of the hydrogen atom. b) Equivalent reduced particle with reduced mass μ at distance r from center of mass.

Since the internal motion of any two-particle system can be represented by the motion of a single particle with a reduced mass, the description of the hydrogen atom has much in common with the description of a diatomic molecule. The Schrödinger Equation for the hydrogen atom.

$$\hat{H}(r,\theta,\varphi)\psi(r,\theta,\varphi) = E\psi(r,\theta,\varphi)$$

employs the same kinetic energy operator, $\hat{T}$, written in spherical coordinates. For the hydrogen atom, however, the distance, r, between the two particles can vary, unlike the diatomic molecule where the bond length was fixed, and the rigid rotor model was used. The hydrogen atom Hamiltonian also contains a potential energy term, $\hat{V}$, to describe the attraction between the proton and the electron. This term is the Coulomb potential energy:

$$\hat{V}(r) - \frac{e^2}{4\pi \in_0 r},$$

where r is the distance between the electron and the proton. The Coulomb potential energy depends inversely on the distance between the electron and the nucleus and does not depend on any angles. Such a potential is called a central potential.

It is convenient to switch from Cartesian coordinates x,y,z to spherical coordinates in terms of a radius r, as well as angles ϕ, which is measured from the positive x axis in the xy plane and may be between 0 and 2π, and θ, which is measured from the positive z axis towards the xy plane and may be between 0 and π.

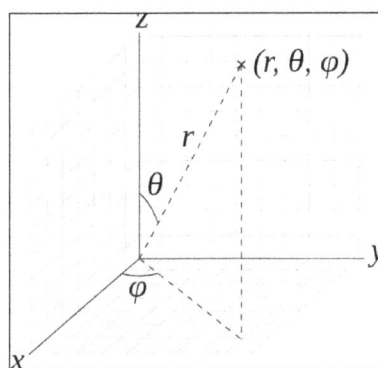

Spherical Coordinates.

The time-independent Schrödinger equation (in spherical coordinates) for a electron around a positively charged nucleus is then.

$$\left\{ -\frac{\hbar^2}{2\mu r^2} \left[\frac{\partial}{\partial r}\left(r^2 \frac{\partial}{\partial r}\right) + \frac{1}{\sin\theta}\frac{\partial}{\partial\theta}\left(\sin\theta\frac{\partial}{\partial\theta}\right) + \frac{1}{\sin^2\theta}\frac{\partial^2}{\partial\varphi^2} \right] - \frac{e^2}{4\pi\epsilon_0 r} \right\} \psi(r,\theta,\varphi) = E\psi(r,\theta,\varphi)$$

Since the angular momentum operator does not involve the radial variable, r, we can separate variables in Equation by using a product wavefunction. We know that the eigenfunctions of the angular momentum operator are the Spherical Harmonic functions, $Y(\theta,\varphi)$, so a good choice for a product function is:

$$\psi(r,\theta,\varphi) = R(r)Y(\theta,\varphi)$$

The Spherical Harmonic $Y(\theta,\phi)$ functions provide information about where the electron is around the proton, and the radial function $R(r)$ describes how far the electron is away from the proton. A solution for both $R(r)$ and $Y(\theta,\phi)$ with E_n that depends on only one quantum number n, although others are required for the proper description of the wavefunction,

$$E_n = -\frac{m_e e^4}{8\epsilon_0^2 h^2 n^2}$$

with $n = 1, 2, 3...\infty$

The hydrogen atom wavefunctions, $\psi(r,\theta,\phi)$, are called atomic orbitals. An atomic orbital is a function that describes one electron in an atom. The wavefunction with $n = 1, l$ $l = 0$ is called the 1s orbital, and an electron that is described by this function is said to be "in" the 1s orbital, i.e. have a 1s orbital state. The constraints on n, l l), and ml that are imposed during the solution of the hydrogen atom Schrödinger equation explain why there is a single 1s orbital, why there are three 2p orbitals, five 3d orbitals, etc. We will see when we consider multi-electron atoms, these constraints explain the features of the Periodic Table. In other words, the Periodic Table is a manifestation of the Schrödinger model and the physical constraints imposed to obtain the solutions to the Schrödinger equation for the hydrogen atom.

The Three Quantum Numbers

Schrödinger's approach requires three quantum numbers (n, l, and ml) to specify a wavefunction for the electron. The quantum numbers provide information about the spatial distribution of an electron. Although n can be any positive integer (NOT zero), only certain values of l and ml are allowed for a given value of (n).

The principal quantum number (n): One of three quantum numbers that tells the average relative distance of an electron from the nucleus. indicates the energy of the electron and the average distance of an electron from the nucleus.

$$n = 1, 2, 3, 4, ...$$

Asn increases for a given atom, so does the average distance of an electron from the nucleus. A negatively charged electron that is, on average, closer to the positively charged nucleus is attracted to the nucleus more strongly than an electron that is farther out in space. This means that electrons with higher values of n are easier to remove from an atom. All wave functions that have the same value of n are said to constitute a principal shell. All the wave functions that have the same value of n because those electrons have similar average distances from the nucleus. because those electrons have similar average distances from the nucleus. The principal quantum number n corresponds to the n used by Bohr to describe electron orbits and by Rydberg to describe atomic energy levels.

The Azimuthal Quantum Number: The second quantum number is often called the azimuthal quantum number (l). One of three quantum numbers that describes the shape of the region of space occupied by an electron. The value of l describes the shape of the region of space occupied by the electron. The allowed values of l depend on the value of n and can range from 0 to n − 1:

$$l = 0, 1, ., 2, 3, .. (n-1)$$

For example, if $n = 1$, l can be only 0; if $n = 2$, l can be 0 or 1; and so forth. For a given atom, all wave functions that have the same values of both n and l form a subshell. A group of wave functions that have the same values of n and l. The regions of space occupied by electrons in the same subshell usually have the same shape, but they are oriented differently in space.

The Magnetic Quantum Number: The third quantum number is the magnetic quantum number (ml). One of three quantum numbers that describes the orientation of the region of space occupied by an electron with respect to an applied magnetic field. The value of ml describes the orientation of the region in space occupied by an electron with respect to an applied magnetic field. The allowed values of ml depend on the value of l: ml can range from –l to l in integral steps:

$$m = -l, -l+1, \ldots 0, \ldots l-1, l$$

For example, if l=0, ml can be only 0; if l = 1, ml can be –1, 0, or +1; and if l = 2, ml can be –2, –1, 0, +1, or +2.

Each wave function with an allowed combination of n, l, and ml values describes an atomic orbital A wave function with an allowed combination of n, l and ml quantum numbers, a particular spatial distribution for an electron. For a given set of quantum numbers, each principal shell has a fixed number of subshells, and each subshell has a fixed number of orbitals.

Rather than specifying all the values of n and l every time we refer to a subshell or an orbital, chemists use an abbreviated system with lowercase letters to denote the value of l for a particular subshell or orbital:

l =	0	1	2	3
Designation	s	p	d	f

The principal quantum number is named first, followed by the letter s, p, d, or f as appropriate. These orbital designations are derived from corresponding spectroscopic characteristics of lines involving them: sharp, principle, diffuse, and fundamental. A 1s orbital has $n = 1$ and l = 0; a 2p subshell has $n = 2$ and $l = 1$ (and has three 2p orbitals, corresponding to $ml = -1$, 0, and +1); a 3d subshell has $n = 3$ and $l = 2$ (and has five 3d orbitals, corresponding to $ml = -2$, –1, 0, +1, and +2); and so forth.

We can summarize the relationships between the quantum numbers and the number of subshells and orbitals as follows:

- Each principal shell has n subshells. For $n = 1$, only a single subshell is possible (1s); for $n = 2$, there are two subshells (2s and 2p); for $n = 3$, there are three subshells (3s, 3p, and 3d); and so forth. Every shell has an ns subshell, any shell with $n \geq 2$ also has an np subshell, and any shell with $n \geq 3$ also has an nd subshell. Because a 2d subshell would require both $n = 2$ and l = 2, which is not an allowed value of l for $n = 2$, a 2d subshell does not exist.

- Each subshell has 2l + 1 orbitals. This means that all ns subshells contain a single s orbital, all np subshells contain three p orbitals, all nd subshells contain five d orbitals, and all nf subshells contain seven f orbitals.

Table: Allowed values of n, l, and ml through n = 4

n	l	Subshell Designation	m_l	Number of Orbitals in Subshell	Number of Orbitals in Shell
1	0	1s	0	1	1
2	0	2s	0	1	4
	1	2p	−1, 0, 1	3	
3	0	3s	0	1	9
	1	3p	−1, 0, 1	3	
	2	3d	−2, −1, 0, 1, 2	5	
4	0	4s	0	1	16
	1	4p	−1, 0, 1	3	
	2	4d	−2, −1, 0, 1, 2	5	
	3	4f	−3, −2, −1, 0, 1, 2, 3	7	

The Radial Component

The first six radial functions are provided in Table. Note that the functions in the table exhibit a dependence on Z, the atomic number of the nucleus. Other one electron systems have electronic states analogous to those for the hydrogen atom, and inclusion of the charge on the nucleus allows the same wavefunctions to be used for all one-electron systems. For hydrogen, $Z=1$ and for helium, $Z=2$.

Table: Hydrogen-like atomic wavefunctions for n values 1,2,3: Z is the atomic number of the nucleus, and $\rho = \dfrac{Zr}{a_0}$ where ao is the Bohr radius and r is the radial variable.

n	ℓ	m	Radial Component
n=1	ℓ=0	m=0	$\psi_{100} = \dfrac{1}{\sqrt{\pi}} - \left(\dfrac{Z}{a_0}\right)^{\frac{3}{2}} e^{-\rho}$
n=2	ℓ=0	m=0	$\psi_{200} = \dfrac{1}{\sqrt{32\pi}} - \left(\dfrac{Z}{a_0}\right)^{\frac{3}{2}} (2-\rho)e^{\frac{-\rho}{2}}$
	ℓ=1	m=0	$\psi_{210} = \dfrac{1}{\sqrt{32\pi}} - \left(\dfrac{Z}{a_0}\right)^{\frac{3}{2}} \rho e^{-\rho/2} \cos(\theta)$
	ℓ=1	m=±1	$\psi_{21\pm1} = \dfrac{1}{\sqrt{64\pi}} - \left(\dfrac{Z}{a_0}\right)^{\frac{3}{2}} \rho e^{-\rho/2} \sin(\theta)e^{\pm i\phi}$

n=3	ℓ=0	m=0	$\psi_{300} = \dfrac{1}{81\sqrt{3\pi}} \left(\dfrac{Z}{a_0}\right)^{\frac{3}{2}} (27 - 18\rho + 2\rho^2)e^{-\rho/3}$
	ℓ=1	m=0	$\psi_{310} = \dfrac{1}{81}\sqrt{\dfrac{2}{\pi}} \left(\dfrac{Z}{a_0}\right)^{\frac{3}{2}} (6r - \rho^2)e^{-\rho/3}\cos(\theta)$
	ℓ=1	m=±1	$\psi_{31\pm1} = \dfrac{1}{81\sqrt{\pi}} \left(\dfrac{Z}{a_0}\right)^{\frac{3}{2}} (6\rho - \rho^2)e^{-r/3}\sin(\theta)e^{\pm i\phi}$
	ℓ=2	m=0	$\psi_{320} = \dfrac{1}{81\sqrt{6\pi}} \left(\dfrac{Z}{a_0}\right)^{\frac{3}{2}} \rho^2 e^{-\rho/3}(3\cos^2(\theta) - 1)$
	ℓ=2	m=±1	$\psi_{32\pm1} = \dfrac{1}{81\sqrt{\pi}} \left(\dfrac{Z}{a_0}\right)^{\frac{3}{2}} \rho^2 e^{-\rho/3}\sin(\theta)\cos(\theta)e^{\pm i\phi}$
	ℓ=2	m=±2	$\psi_{32\pm2} = \dfrac{1}{162\sqrt{\pi}} \left(\dfrac{Z}{a_0}\right)^{\frac{3}{2}} \rho^2 e^{-\rho/3}\sin^2(\theta)e^{\pm 2i\phi}$

Visualizing the variation of an electronic wavefunction with r, θ, and φ is important because the absolute square of the wavefunction depicts the charge distribution (electron probability density) in an atom or molecule. The charge distribution is central to chemistry because it is related to chemical reactivity. For example, an electron deficient part of one molecule is attracted to an electron rich region of another molecule, and such interactions play a major role in chemical interactions ranging from substitution and addition reactions to protein folding and the interaction of substrates with enzymes.

Methods for separately examining the radial portions of atomic orbitals provide useful information about the distribution of charge density within the orbitals. Graphs of the radial functions, R(r), for the 1s, 2s, and 2p orbitals plotted in figure left). The quantity R(r)*R(r) gives the radial probability density; i.e., the probability density for the electron to be at a point located the distance r from the proton. Radial probability densities for three types of atomic orbitals are plotted in figure (right).

For the hydrogen atom, the peak in the radial probability plot occurs at r = 0.529 Å (52.9 pm), which is exactly the radius calculated by Bohr for the $n = 1$ orbit. Thus the most probable radius obtained from quantum mechanics is identical to the radius calculated by classical mechanics. In Bohr's model, however, the electron was assumed to be at this distance 100% of the time, whereas in the Schrödinger model, it is at this distance only some of the time. The difference between the two models is attributable to the wavelike behavior of the electron and the Heisenberg uncertainty principle.

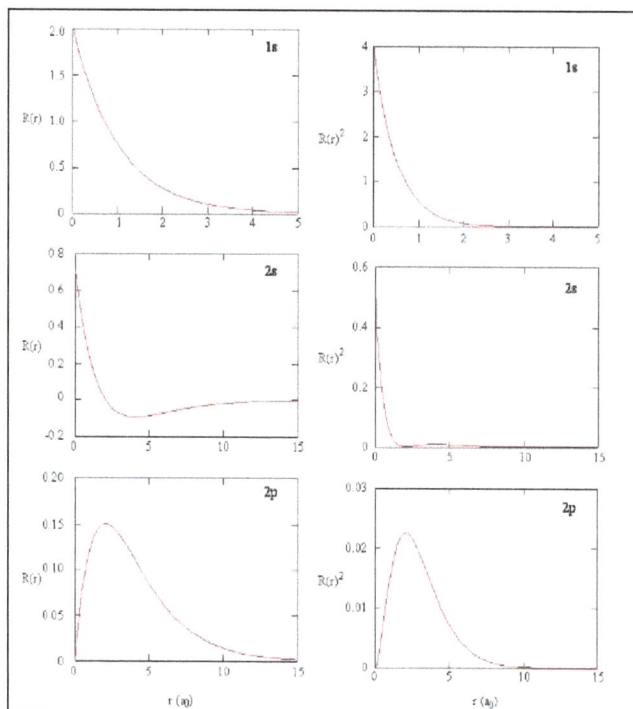

(left) Radial function, R(r), for the 1s, 2s, and 2p orbitals. (right)
Radial probability densities for the 1s, 2s, and 2p orbitals.

Figure compares the electron probability densities for the hydrogen 1s, 2s, and 3s orbitals. Note that all three are spherically symmetrical. For the 2s and 3s orbitals, however (and for all other s orbitals as well), the electron probability density does not fall off smoothly with increasing r. Instead, a series of minima and maxima are observed in the radial probability plots (part (c) in figure). The minima correspond to spherical nodes (regions of zero electron probability), which alternate with spherical regions of nonzero electron probability.

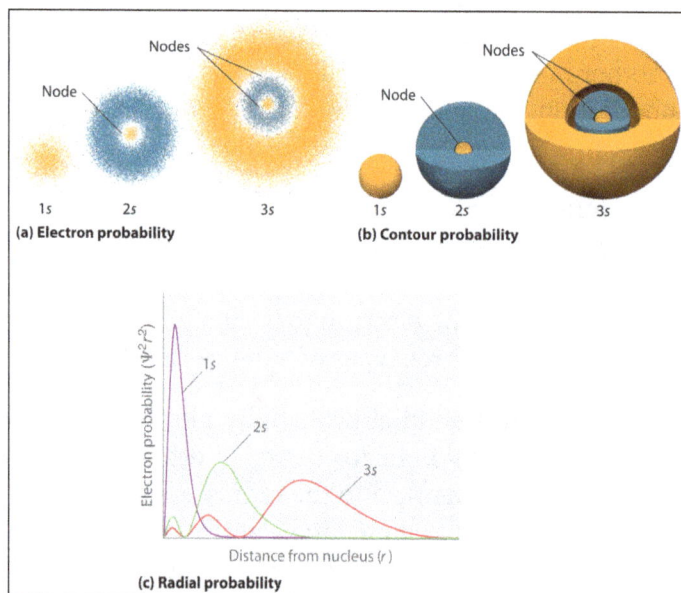

Probability Densities for the 1s, 2s, and 3s Orbitals of the Hydrogen Atom.
(a) The electron probability density in any plane that contains the nucleus is shown.

Note the presence of circular regions, or nodes, where the probability density is zero. (b) Contour surfaces enclose 90% of the electron probability, which illustrates the different sizes of the 1s, 2s, and 3s orbitals. The cutaway drawings give partial views of the internal spherical nodes. The orange color corresponds to regions of space where the phase of the wave function is positive, and the blue color corresponds to regions of space where the phase of the wave function is negative. (c) In these plots of electron probability as a function of distance from the nucleus (r) in all directions (radial probability), the most probable radius increases as increases, but the 2s and 3s orbitals have regions of significant electron probability at small values of r.

The Angular Component

The angular component of the wavefunction $Y(\theta,\phi)$ in Equation does much to give an orbital its distinctive shape. $Y(\theta,\phi)$ is typically normalized so the the integral of $Y^2(\theta,\phi)$ over the unit sphere is equal to one. In this case, $Y^2(\theta,\phi)$ serves as a probability function. The probability function can be interpreted as the probability that the electron will be found on the ray emitting from the origin that is at angles (θ,ϕ) from the axes. The probability function can also be interpreted as the probability distribution of the electron being at position (θ,ϕ) on a sphere of radius r, given that it is r distance from the nucleus. $Y_{l,m_l}(\theta,\phi)$ are also the wavefunction solutions to Schrödinger's equation for a rigid rotor consisting of rotating bodies, for example a diatomic molecule. These are called Spherical Harmonic functions.

s Orbitals (l=0)

Three things happen to s orbitals as n increases:

- They become larger, extending farther from the nucleus.

- They contain more nodes. This is similar to a standing wave that has regions of significant amplitude separated by nodes, points with zero amplitude.

- For a given atom, the s orbitals also become higher in energy as n increases because of their increased distance from the nucleus.

Orbitals are generally drawn as three-dimensional surfaces that enclose 90% of the electron density. Although such drawings show the relative sizes of the orbitals, they do not normally show the spherical nodes in the 2s and 3s orbitals because the spherical nodes lie inside the 90% surface. Fortunately, the positions of the spherical nodes are not important for chemical bonding.

p Orbitals (l=1)

Only s orbitals are spherically symmetrical. As the value of l increases, the number of orbitals in a given subshell increases, and the shapes of the orbitals become more complex. Because the 2p subshell has l = 1, with three values of $m_l(-1, 0, \text{ and } +1)$, there are three 2p orbitals).

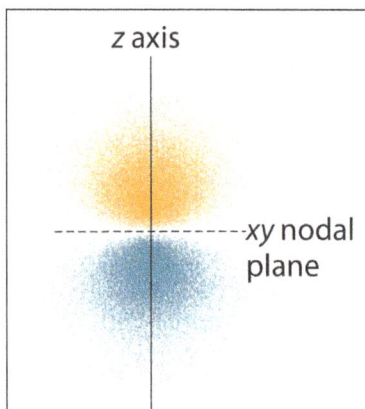

Electron Probability Distribution for a Hydrogen 2p Orbital. The nodal plane of zero electron density separates the two lobes of the 2p orbital. As in figure, the colors correspond to regions of space where the phase of the wave function is positive (orange) and negative (blue).

The electron probability distribution for one of the hydrogen 2p orbitals is shown in figure. Because this orbital has two lobes of electron density arranged along the z axis, with an electron density of zero in the xy plane (i.e., the xy plane is a nodal plane), it is a 2pz orbital. As shown in Figur, the other two 2p orbitals have identical shapes, but they lie along the x axis ($2p_x$) and y axis ($2p_y$), respectively. Note that each p orbital has just one nodal plane. In each case, the phase of the wave function for each of the 2p orbitals is positive for the lobe that points along the positive axis and negative for the lobe that points along the negative axis. It is important to emphasize that these signs correspond to the phase of the wave that describes the electron motion, not to positive or negative charges.

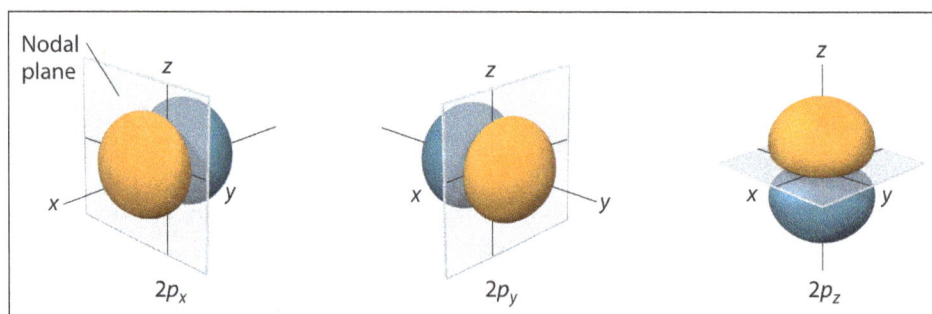

The Three Equivalent 2p Orbitals of the Hydrogen Atom.

The surfaces shown enclose 90% of the total electron probability for the 2px, 2py, and 2pz orbitals. Each orbital is oriented along the axis indicated by the subscript and a nodal plane that is perpendicular to that axis bisects each 2p orbital. The phase of the wave function is positive (orange) in the region of space where x, y, or z is positive and negative (blue) where x, y, or z is negative. Just as with the s orbitals, the size and complexity of the p orbitals for any atom increase as the principal quantum number n increases. The shapes of the 90% probability surfaces of the 3p, 4p, and higher-energy p orbitals are, however, essentially the same as those shown in figures.

d Orbitals (l=2)

Subshells with $l = 2$ have five d orbitals; the first principal shell to have a d subshell corresponds to $n = 3$. The five d orbitals have m_l values of −2, −1, 0, +1, and +2.

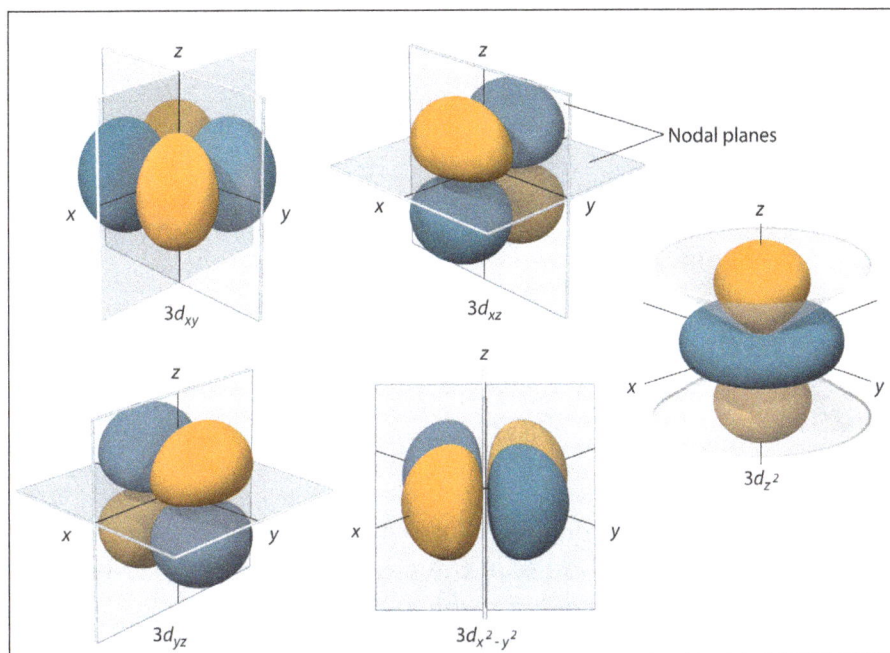

The Five Equivalent 3d Orbitals of the Hydrogen Atom.

The surfaces shown enclose 90% of the total electron probability for the five hydrogen 3d orbitals. Four of the five 3d orbitals consist of four lobes arranged in a plane that is intersected by two perpendicular nodal planes. These four orbitals have the same shape but different orientations. The fifth 3d orbital, $3d_{z^2}$, has a distinct shape even though it is mathematically equivalent to the others. The phase of the wave function for the different lobes is indicated by color: orange for positive and blue for negative.

The hydrogen 3d orbitals have more complex shapes than the 2p orbitals. All five 3d orbitals contain two nodal surfaces, as compared to one for each p orbital and zero for each s orbital. In three of the d orbitals, the lobes of electron density are oriented between the x and y, x and z, and y and z planes; these orbitals are referred to as the $3d_{xy}$, \)3d_{x z}_, and \)3d_{y z}\) orbitals, respectively. A fourth d orbital has lobes lying along the x and y axes; this is the 3dx2−y2 orbital. The fifth 3d orbital, called the $3d_{z^2}$ orbital, has a unique shape: it looks like a $2p_z$ orbital combined with an additional doughnut of electron probability lying in the xy plane. Despite its peculiar shape, the $3d_{z^2}$ orbital is mathematically equivalent to the other four and has the same energy. In contrast to p orbitals, the phase of the wave function for d orbitals is the same for opposite pairs of lobes. As shown in figure the phase of the wave function is positive for the two lobes of the dz² orbital that lie along the z axis, whereas the phase of the wave function is negative for the doughnut of electron density in the xy plane. Like the s and p orbitals, as n increases, the size of the d orbitals increases, but the overall shapes remain similar to those depicted in figure.

f Orbitals (l=3)

Principal shells with $n = 4$ can have subshells with $l = 3$ and m_l values of −3, −2, −1, 0, +1, +2, and +3. These subshells consist of seven *f* orbitals. Each *f* orbital has three nodal surfaces, so their shapes are complex (not shown).

Energies

The constraint that n be greater than or equal to $l+1$ also turns out to quantize the energy, producing the same quantized expression for hydrogen atom energy levels that was obtained from the Bohr model of the hydrogen atom.

$$E = -\frac{Z^2}{n^2} Rhc$$

Or,

$$E_n = -\frac{Z^2 \mu e^4}{8\epsilon_0^2 h^2 n^2}$$

The relative energies of the atomic orbitals with $n \leq 4$ for a hydrogen atom are plotted in figure note that the orbital energies depend on *only* the principal quantum number n. Consequently, the energies of the 2s and 2p orbitals of hydrogen are the same; the energies of the 3s, 3p, and 3d orbitals are the same; and so forth. The orbital energies obtained for hydrogen using quantum mechanics are exactly the same as the allowed energies calculated by Bohr. In contrast to Bohr's model, however, which allowed only one orbit for each energy level, quantum mechanics predicts that there are *4* orbitals with different electron density distributions in the $n = 2$ principal shell (one 2s and three 2p orbitals), *9* in the $n = 3$ principal shell, and *16* in the $n = 4$ principal shell.

The different values of l and m_l for the individual orbitals within a given principal shell are not important for understanding the emission or absorption spectra of the hydrogen atom under most conditions, but they do explain the splittings of the main lines that are observed when hydrogen atoms are placed in a magnetic field. As we have just seen, however, quantum mechanics also predicts that in the hydrogen atom, all orbitals with the same value of n (e.g., the three 2p orbitals) are degenerate, meaning that they have the same energy. Figure shows that the energy levels become closer and closer together as the value of n increases, as expected because of the $1/n^2$ dependence of orbital energies.

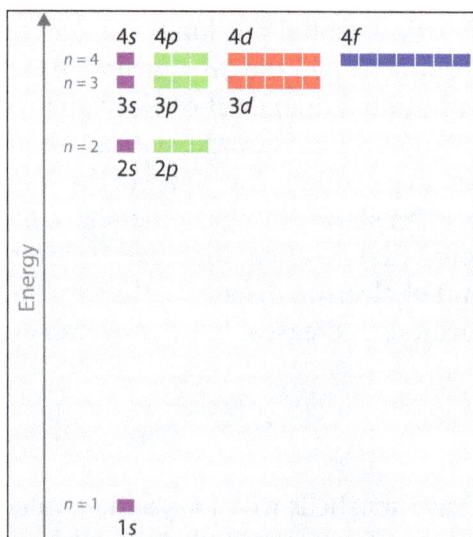

Orbital Energy Level Diagram for the Hydrogen Atom. Each box corresponds to one orbital.
Note that the difference in energy between orbitals decreases rapidly with increasing values of n.

In general, both energy and radius decrease as the nuclear charge increases. Thus the most stable orbitals (those with the lowest energy) are those closest to the nucleus. For example, in the ground state of the hydrogen atom, the single electron is in the 1s orbital, whereas in the first excited state, the atom has absorbed energy and the electron has been promoted to one of the $n = 2$ orbitals. In ions with only a single electron, the energy of a given orbital depends on only n, and all subshells within a principal shell, such as the px, py, and pz orbitals, are degenerate.

Electron Spin: The Fourth Quantum Number

The quantum numbers n, l, m are not sufficient to fully characterize the physical state of the electrons in an atom. In 1926, Otto Stern and Walther Gerlach carried out an experiment that could not be explained in terms of the three quantum numbers n, l, m and showed that there is, in fact, another quantum-mechanical degree of freedom that needs to be included in the theory. The experiment is illustrated in the figure. A beam of atoms (e.g. hydrogen or silver atoms) is sent through a spatially inhomogeneous magnetic field with a definite field gradient toward one of the poles. It is observed that the beam splits into two beams as it passes through the field region.

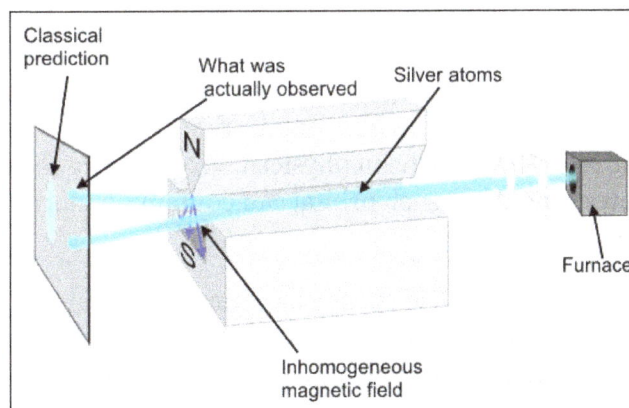

The Stern-Gerlach apparatus.

The fact that the beam splits into 2 beams suggests that the electrons in the atoms have a degree of freedom capable of coupling to the magnetic field. That is, an electron has an intrinsic magnetic moment MM arising from a degree of freedom that has no classical analog. The magnetic moment must take on only 2 values according to the Stern-Gerlach experiment. The intrinsic property that gives rise to the magnetic moment must have some analog to a spin, S; unlike position and momentum, which have clear classical analogs, spin does not. The implication of the Stern-Gerlach experiment is that we need to include a fourth quantum number, msms in our description of the physical state of the electron. That is, in addition to give its principle, angular, and magnetic quantum numbers, we also need to say if it is a spin-up electron or a spin-down electron.

George Uhlenbeck and Samuel Goudsmit, proposed that the splittings were caused by an electron spinning about its axis, much as Earth spins about its axis. When an electrically charged object spins, it produces a magnetic moment parallel to the axis of rotation, making it behave like a magnet. Although the electron cannot be viewed solely as a particle, spinning or otherwise, it is indisputable that it does have a magnetic moment. This magnetic moment is called electron spin.

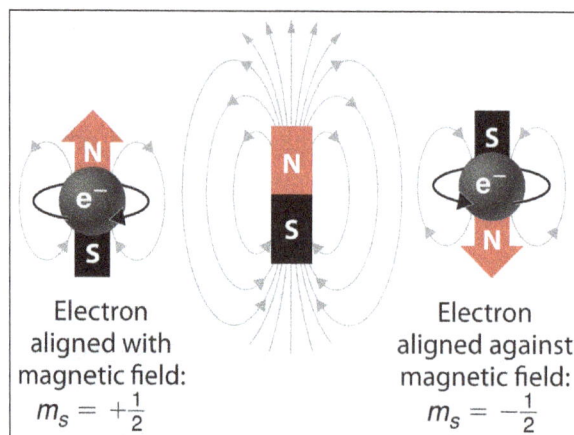

Electron Spin. In a magnetic field, an electron has two possible orientations with
different energies, one with spin up, aligned with the magnetic field, and one
with spin down, aligned against it. All other orientations are forbidden.

In an external magnetic field, the electron has two possible orientations. These are described by
a fourth quantum number (m_s), which for any electron can have only two possible values, designated +½ (up) and −½ (down) to indicate that the two orientations are opposites; the subscript
s is for spin. An electron behaves like a magnet that has one of two possible orientations, aligned
either with the magnetic field or against it. The implications of electron spin for chemistry were
recognized almost immediately by an Austrian physicist, Wolfgang Pauli (1900–1958; Nobel Prize
in Physics, 1945), who determined that each orbital can contain no more than two electrons whoe
developed the Pauli exclusion principle.

By giving the values of n, l, and ml, we also specify a particular orbital (e.g., 1s with n = 1, / = 0, ml
= 0). Because m_s has only two possible values (+½ or −½), two electrons, and only two electrons,
can occupy any given orbital, one with spin up and one with spin down. With this information, we
can proceed to construct the entire periodic table, which was originally based on the physical and
chemical properties of the known elements.

References

- Tsuneda, Takao (2014). "Ch. 3: Electron Correlation". Density Functional Theory in Quantum Chemistry. To-
kyo: Springer. Pp. 73–75. Doi:10.1007/978-4-431-54825-6. ISBN 978-4-431-54825-6.

- Thalau, Peter; Ritz, Thorsten; Burda, Hynek; Wegner, Regina E.; Wiltschko, Roswitha (18 April 2006). "The
magnetic compass mechanisms of birds and rodents are based on different physical principles". Interface [on-
line journal of the Royal Society of London]. 3 (9). PMC 1664646

- Griffiths, David J. (2004). Introduction to Quantum Mechanics(2nd ed.). Prentice Hall. P. 247. ISBN 0-13-
111892-7. OCLC 40251748

- Abdulsattar, Mudar A. (2012). "sige superlattice nanocrystal infrared and Raman spectra: A density functional
theory study". J. Appl. Phys. 111 (4): 044306. Bibcode:2012JAP...111d4306a. Doi:10.1063/1.3686610

- Hinchliffe, Alan (2000). Modelling Molecular Structures (2nd ed.). Baffins Lane, Chichester, West Sussex
PO19 1UD, England: John Wiley & Sons Ltd. P. 186. ISBN 0-471-48993-X

8

Computational Quantum Chemistry

Computational quantum chemistry is one of the fields of chemistry that makes use of computer programs and approximations for solving chemical problems. It calculates electronic charge density, dipoles and multiple moments, absolute and relative energies, vibrational frequencies of molecules and solids. This chapter discusses this field of computational quantum chemistry in detail.

Computational quantum chemistry focuses specifically on equations and approximations derived from the postulates of quantum mechanic.

Quantum Calculations

Multielectron Electronic Wavefunctions

We could symbolically write an approximate two-particle wavefunction as $\psi(r_1, r_2)$. This could be, for example, a two-electron wavefunction for helium. To exchange the two particles, we simply substitute the coordinates of particle 1 (r_1) for the coordinates of particle 2 (r_2) and vice versa, to get the new wavefunction $\psi(r_1, r_2)$. This new wavefunction must have the property that:

$$|\psi(r_1, r_2)|^2 = \psi(r_2, r_1)^* \psi(r_2, r_1) = \psi(r_1, r_2)^* \psi(r_1, r_2)$$

Equation $|\psi(r_1, r_2)|^2 = \psi(r_2, r_1)^* \psi(r_2, r_1) = \psi(r_1, r_2)^* \psi(r_1, r_2)$ will be true only if the wavefunctions before and after permutation are related by a factor of $e^{i\varphi}$:

$$\psi(r_1, r_2) = e^{i\varphi} \psi(r_1, r_2)$$

so that,

$$(e^{-i\varphi} \psi(r_1, r_2)^*)(e^{i\varphi} \psi(r_1, r_2)^*) = \psi(r_1, r_2)^* \psi(r_1, r_2)$$

If we exchange or permute two identical particles twice, we are (by definition) back to the original situation. If each permutation changes the wavefunction by $e^{i\varphi}$, the double permutation must change the wavefunction by $e^{i\varphi} e^{i\varphi}$. Since we then are back to the original state, the effect of the double permutation must equal 1; i.e.,

$$e^{i\varphi}e^{i\varphi} = e^{i2\varphi} = 1$$

which is true only if $\varphi = 0$ or an integer multiple of π. The requirement that a double permutation reproduce the original situation limits the acceptable values for $e^{i\varphi}$ to either +1 (when $\varphi = 0$) or -1 (when $\varphi = \delta$). Both possibilities are found in nature, but the behavior of elections is that the wavefunction be antisymmetric with respect to permutation ($e^{i\varphi} = -1$). A wavefunction that is anti-symmetric with respect to electron interchange is one whose output changes sign when the electron coordinates are interchanged, as shown below.

$$\psi(r_2, r_1) = e \ \psi(r_1, r_2) = -\psi(r_1, r_2)$$

Blindly following the first statement of the Pauli Exclusion Principle, that each electron in a multi-electron atom must be described by a different spin-orbital, we try constructing a simple product wavefunction for helium using two different spin-orbitals. Both have the 1s spatial component, but one has spin function $\alpha\alpha$ and the other has spin function βso the product wavefunction matches the form of the ground state electron configuration for He,1s².

$$\psi(r_1, r_2) = \varphi_{1s\alpha}(r_1)\varphi_{1s\beta}(r_2)$$

After permutation of the electrons, this becomes:

$$\psi(r_2, r_1) = \varphi_{1s\alpha}(r_2)\varphi_{1s\beta}(r_1)$$

Which is different from the starting function since $\varphi_{1s\alpha}$ and $\varphi_{1s\beta}$ are different spin-orbital func-tions. However, an antisymmetric function must produce the same function multiplied by (−1) after permutation, and that is not the case here.

To avoid getting a totally different function when we permute the electrons, we can make a linear combination of functions. A very simple way of taking a linear combination involves making a new function by simply adding or subtracting functions. The function that is created by subtracting the right-hand side of Equation $\psi(r_2, r_1) = \varphi_{1s\alpha}(r_2)\varphi_{1s\beta}(r_1)$ from the right-hand side of Equation $\psi(r_1, r_2) = \varphi_{1s\alpha}(r_1)\varphi_{1s\beta}(r_2)$ has the desired antisymmetric behavior. The constant on the right-hand side accounts for the fact that the total wavefunction must be normalized.

$$\psi(r_1, r_2) = \frac{1}{\sqrt{2}} - [\varphi_{1s\alpha}(r_1)\varphi_{1s\beta}(r_2) - \varphi_{1s\alpha}(r_2)\varphi_{1s\beta}(r_1)]$$

A linear combination that describes an appropriately antisymmetrized multi-electron wavefunc-tion for any desired orbital configuration is easy to construct for a two-electron system. However, interesting chemical systems usually contain more than two electrons. For these multi-electron systems a relatively simple scheme for constructing an antisymmetric wavefunction from a prod-uct of one-electron functions is to write the wavefunction in the form of a determinant. John Slater introduced this idea so the determinant is called a Slater determinant.

The Slater determinant for the two-electron wavefunction for the ground state H_2 system (with the two electrons occupying the σ_{1s} molecular orbital),

$$\psi(r_1, r_2) = \frac{1}{\sqrt{2}} - \begin{vmatrix} \sigma_{1s}(1)\alpha(1) & \sigma_{1s}(1)\beta(1) \\ \sigma_{1s}(2)\alpha(2) & \sigma_{1s}(2)\beta(2) \end{vmatrix}$$

We can introduce a shorthand notation for the arbitrary spin-orbital:

$$\chi_{i\alpha}(r) = \varphi_i \alpha$$

or,

$$\chi_{i\beta}(r) = \varphi_i \beta$$

as determined by the msms quantum number. A shorthand notation for the determinant in Equation is then:

$$\psi(r_1, r_2) = 2^{-\frac{1}{2}} Det \,|\, \chi_{1s\alpha}(r_1)\alpha\chi_{1s\beta}(r_2)\beta \,|$$

The determinant is written so the electron coordinate changes in going from one row to the next, and the spin orbital changes in going from one column to the next. The advantage of having this recipe is clear if you try to construct an antisymmetric wavefunction that describes the orbital configuration for uranium! Note that the normalization constant is:

$$(N!) - \frac{1}{2}$$

for a system of N electrons.

The generalized Slater determinant for a multe-electrom atom with N electrons is then:

$$\psi(r_1, r_2, \ldots, r_N) = \frac{1}{\sqrt{N!}} \begin{vmatrix} \chi_1(r_1)\alpha & \chi_1(r_1)\beta & \cdots & \chi_{N/2}(r_1)\beta \\ \chi_1(r_2)\alpha & \chi_2(r_2)\beta & \cdots & \chi_{N/2}(r_2)\beta \\ \vdots & \vdots & \ddots & \vdots \\ \chi_1(r_N)\alpha & \chi_2(r_N)\beta & \cdots & \chi_{N/2}(r_N)\beta \end{vmatrix}$$

In a modern ab initio electronic structure calculation on a closed shell molecule, the electronic Hamiltonian is used with a single determinant wavefunction. This wavefunction, Ψ, is constructed from molecular orbitals, ψ that are written as linear combinations of contracted Gaussian basis functions, φ $\varphi_j = \sum_k c_{jk}\psi_k$.

The contracted Gaussian functions are composed from primitive Gaussian functions to match Slater-type orbitals. The exponential parameters in the STOs are optimized by calculations on small molecules using the nonlinear variational method and then those values are used with other molecules. The problem is to calculate the electronic energy from:

$$E = \frac{\int \Psi^* \hat{H} \Psi \, d\tau}{\int \Psi * \Psi \, d\tau}$$

or in bra-ket notation:

$$E = \frac{\left\langle \Psi \mid \hat{H} \mid \Psi \right\rangle}{\left\langle \psi \mid \psi \right\rangle}$$

The the optimum coefficients c_{jk} for each molecular orbital in Equation $\varphi_j = \sum c_{jk} \psi_k$ by using the Self Consistent Field Method and the Linear Variational Method to minimize the energy as was described previously for atoms.

The variational principle says an approximate energy is an upper bound to the exact energy, so the lowest energy that we calculate is the most accurate. At some point, the improvements in the energy will be very slight. This limiting energy is the lowest that can be obtained with a single determinant wavefunction (e.g., Equation $\psi(r_1, r_2, \ldots, r_N) = \dfrac{1}{\sqrt{N!}} \begin{vmatrix} \chi_1(r_1)\alpha & \chi_1(r_1)\beta & \cdots & \chi_{N/2}(r_1)\beta \\ \chi_1(r_2)\alpha & \chi_2(r_2)\beta & \cdots & \chi_{N/2}(r_2)\beta \\ \vdots & \vdots & \ddots & \vdots \\ \chi_1(r_N)\alpha & \chi_2(r_N)\beta & \cdots & \chi_{N/2}(r_N)\beta \end{vmatrix}$).

This limit is called the Hartree-Fock limit, the energy is the Hartree-Fock energy, the molecular orbitals producing this limit are called Hartree-Fock orbitals, and the determinant is the Hartree-Fock wavefunction.

Hartree-Fock Calculations

You may encounter the terms *restricted* and *unrestricted* Hartree-Fock. In a restricted HF calculation, electrons with $\alpha\alpha$ spin are restricted or constrained to occupy the same spatial orbitals as electrons with β spin. This constraint is removed in an unrestricted calculation. For example, the spin orbital for electron 1 could be $\psi_A(r_1)\alpha(1)$ and the spin orbital for electron 2 in a molecule could be $\psi_A(r_2)\alpha(2)$, where both the spatial molecular orbital and the spin function differ for the two electrons. Such spin orbitals are called *unrestricted*. If both electrons are constrained to have the same spatial orbital, e.g. $\psi_A(r_1)\alpha(1)$ and $\psi_A(r_2)\alpha(2)$, then the spin orbital is said to be *restricted*. While unrestricted spin orbitals can provide a better description of the electrons, twice as many spatial orbitals are needed, so the demands of the calculation are much higher. Using unrestricted orbitals is particular beneficial when a molecule contains an odd number of electrons because there are more electrons in one spin state than in the other.

Carbon Monoxide

It is well known that carbon monoxide is a poison that acts by binding to the iron in hemoglobin and preventing oxygen from binding. As a result, oxygen is not transported by the blood to cells. Which end of carbon monoxide, carbon or oxygen, do you think binds to iron by donating electrons? We all know that oxygen is more electron-rich than carbon (8 vs 6 electrons) and more electronegative. A reasonable answer to this question therefore is oxygen, but experimentally it is carbon that binds to iron.

A quantum mechanical calculation done by Winifred M. Huo, published in J. Chem. Phys. 43, 624, provides an explanation for this counter-intuitive result. The basis set used in the calculation consisted of 10 functions: the 1s, 2s, $2p_x$, $2p_y$, and $2p_z$ atomic orbitals of C and O. Ten molecular orbitals (mo's) were defined as linear combinations of the ten atomic orbitals (Equation $\varphi_j = \sum_k c_{jk}\psi_k$

. The ground state wavefunction $\Psi\Psi$ is written as the Slater Determinant of the five lowest energy

molecular orbitals   . Equation $E = \dfrac{\int \Psi^* \hat{H} \Psi \, d\tau}{\int \Psi * \Psi \, d\tau}$ gives the energy of the ground state, where the

denominator accounts for the normalization requirement. The coefficients c_{kj} in the linear combination are determined by the variational method to minimize the energy. The solution of this problem gives the following equations for the molecular orbitals. Only the largest terms have been retained here. These functions are listed and discussed in order of increasing energy.

- $1s \approx 0.941s_o$. The 1 says this is the first σ orbital. The σ says it is symmetric with respect to reflection in the plane of the molecule. The large coefficient, 0.94, means this is essentially the 1s atomic orbital of oxygen. The oxygen 1s orbital should have a lower energy than that of carbon because the positive charge on the oxygen nucleus is greater.

- $2s \approx 0.921s_c$. This orbital is essentially the 1s atomic orbital of carbon. Both the σ1σ and 2σ2σ are "nonbonding" orbitals since they are localized on a particular atom and do not directly determine the charge density between atoms.

- $3s \approx \left(0.722s_o + 0.182p_{zo}\right) + \left(0.282s_c + 0.162p_{zc}\right)$. This orbital is a "bonding" molecular orbital because the electrons are delocalized over C and O in a way that enhances the charge density between the atoms. The 3 means this is the third σ orbital. This orbital also illustrates the concept of hybridization. One can say the 2s and 2p orbitals on each atom are hybridized and the molecular orbital is formed from these hybrids although the calculation just obtains the linear combination of the four orbitals directly without the à prioriintroduction of hybridization. In other words, hybridization just falls out of the calculation. The hybridization in this bonding LCAO increases the amplitude of the function in the region of space between the two atoms and decreases it in the region of space outside of the bonding region of the atoms.

- $4s \approx (0.372s_c + 0.12p_{zc}) + (0.542p_{zo} - 0.432s_o)$. This molecular orbital also can be thought of as being a hybrid formed from atomic orbitals. The hybridization of oxygen atomic orbitals, because of the negative coefficient with $2s_O$, decreases the electron density between the nuclei and enhances electron density on the side of oxygen facing away from the carbon atom. If we follow how this function varies along the internuclear axis, we see that near carbon the function is positive whereas near oxygen it is negative or possibly small and positive. This change means there must be a node between the two nuclei or at the oxygen nucleus. Because of the node, the electron density between the two nuclei is low so the electrons in this orbital do not serve to shield the two positive nuclei from each other. This orbital therefore is called an "antibonding" molecular orbital and the electrons assigned to it are called antibonding electrons. This orbital is the antibonding partner to the 3σ orbital.

- $1\pi \approx 0.322p_{xc} + 0.442p_{xo}$ and $2\pi \approx 0.322p_{yc} + 0.442p_{yo}$. These two orbitals are degenerate and correspond to bonding orbitals made up from the p_x and p_y atomic orbitals from each atom. These orbitals are degenerate because the x and y directions are equivalent in this molecule. $\pi\pi$ tells us that these orbitals are antisymmetric with respect to reflection in a plane containing the nuclei.

- $5\sigma \approx 0.382_{sC} - 0.382_{pC} - 0.292 p_{zO}$. This orbital is the sp hybrid of the carbon atomic orbitals. The negative coefficient for $2p_C$ puts the largest amplitude on the side of carbon away from oxygen. There is no node between the atoms. We conclude this is a nonbonding orbital with the nonbonding electrons on carbon. This is not a "bonding" orbital because the electron density between the nuclei is lowered by hybridization. It also is not an antibonding orbital because there is no node between the nuclei. When carbon monoxide binds to Fe in hemoglobin, the bond is made between the C and the Fe. This bond involves the donation of the $5\sigma5\sigma$ nonbonding electrons on C to empty d orbitals on Fe. Thus molecular orbital theory allows us to understand why the C end of the molecule is involved in this electron donation when we might naively expect O to be more electron-rich and capable of donating electrons to iron.

Basis Set

A basis set in theoretical and computational chemistry is a set of functions (called basis functions) that is used to represent the electronic wave function in the Hartree–Fock method or density-functional theory in order to turn the partial differential equations of the model into algebraic equations suitable for efficient implementation on a computer.

The use of basis sets is equivalent to the use of an approximate resolution of the identity. The single-particle states (molecular orbitals) are then expressed as linear combinations of the basis functions.

The basis set can either be composed of atomic orbitals (yielding the linear combination of atomic orbitals approach), which is the usual choice within the quantum chemistry community, or plane waves which are typically used within the solid state community. Several types of atomic orbitals can be used: Gaussian-type orbitals, Slater-type orbitals, or numerical atomic orbitals. Out of the three, Gaussian-type orbitals are by far the most often used, as they allow efficient implementations of Post-Hartree–Fock methods.

In modern computational chemistry, quantum chemical calculations are performed using a finite set of basis functions. When the finite basis is expanded towards an (infinite) complete set of functions, calculations using such a basis set are said to approach the complete basis set (CBS) limit. basis function and atomic orbital are sometimes used interchangeably, although the basis functions are usually not true atomic orbitals, because many basis functions are used to describe polarization effects in molecules.

Within the basis set, the wavefunction is represented as a vector, the components of which correspond to coefficients of the basis functions in the linear expansion. In such a basis, one-electron operators correspond to matrices (a.k.a. rank two tensors), whereas two-electron operators are rank four tensors.

When molecular calculations are performed, it is common to use a basis composed of atomic orbitals, centered at each nucleus within the molecule (linear combination of atomic orbitals ansatz).

The physically best motivated basis set are Slater-type orbitals (STOs), which are solutions to the Schrödinger equation of hydrogen-like atoms, and decay exponentially far away from the nucleus. It can be shown that the molecular orbitals of Hartree-Fock and density-functional theory also exhibit exponential decay. Furthermore, S-type STOs also satisfy Kato's cusp condition at the nucleus, meaning that they are able to accurately describe electron density near the nucleus. However, hydrogen-like atoms lack many-electron interactions, thus the orbitals do not accurately describe electron state correlations.

Unfortunately, calculating integrals with STOs is computationally difficult and it was later realized by Frank Boys that STOs could be approximated as linear combinations of Gaussian-type orbitals (GTOs) instead. Because the product of two GTOs can be written as a linear combination of GTOs, integrals with Gaussian basis functions can be written in closed form, which leads to huge computational savings.

Dozens of Gaussian-type orbital basis sets have been published in the literature. Basis sets typically come in hierarchies of increasing size, giving a controlled way to obtain more accurate solutions, however at a higher cost.

The smallest basis sets are called minimal basis sets. A minimal basis set is one in which, on each atom in the molecule, a single basis function is used for each orbital in a Hartree–Fock calculation on the free atom. For atoms such as lithium, basis functions of p type are also added to the basis functions that correspond to the 1s and 2s orbitals of the free atom, because lithium also has a 1s2p bound state. For example, each atom in the second period of the periodic system (Li - Ne) would have a basis set of five functions (two s functions and three p functions).

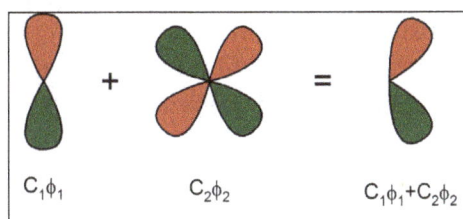

$C_1\phi_1$ $C_2\phi_2$ $C_1\phi_1 + C_2\phi_2$

A d-polarization function added to a p orbital.

The minimal basis set is close to exact for the gas-phase atom. In the next level, additional functions are added to describe polarization of the electron density of the atom in molecules. These are called polarization functions. For example, while the minimal basis set for hydrogen is one function approximating the 1s atomic orbital, a simple polarized basis set typically has two s- and one p-function (which consists of three basis functions: px, py and pz). This adds flexibility to the basis set, effectively allowing molecular orbitals involving the hydrogen atom to be more asymmetric about the hydrogen nucleus. This is very important for modeling chemical bonding, because the bonds are often polarized. Similarly, d-type functions can be added to a basis set with valence p orbitals, and f-functions to a basis set with d-type orbitals, and so on.

Another common addition to basis sets is the addition of diffuse functions. These are extended Gaussian basis functions with a small exponent, which give flexibility to the "tail" portion of the atomic orbitals, far away from the nucleus. Diffuse basis functions are important for describing anions or dipole moments, but they can also be important for accurate modeling of intra- and intermolecular bonding.

Minimal Basis Sets

The most common minimal basis set is STO-nG, where n is an integer. This n value represents the number of Gaussian primitive functions comprising a single basis function. In these basis sets, the same number of Gaussian primitives comprise core and valence orbitals. Minimal basis sets typically give rough results that are insufficient for research-quality publication, but are much cheaper than their larger counterparts. Commonly used minimal basis sets of this type are:

- STO-3G,

- STO-4G,

- STO-6G,

- STO-3G* - Polarized version of STO-3G.

There are several other minimum basis sets that have been used such as the MidiX basis sets.

Split-Valence Basis Sets

During most molecular bonding, it is the valence electrons which principally take part in the bonding. In recognition of this fact, it is common to represent valence orbitals by more than one basis function (each of which can in turn be composed of a fixed linear combination of primitive Gaussian functions). Basis sets in which there are multiple basis functions corresponding to each valence atomic orbital are called valence double, triple, quadruple-zeta, and so on, basis sets (zeta, ζ, was commonly used to represent the exponent of an STO basis function). Since the different orbitals of the split have different spatial extents, the combination allows the electron density to adjust its spatial extent appropriate to the particular molecular environment. In contrast, minimal basis sets lack the flexibility to adjust to different molecular environments.

Pople Basis Sets

The notation for the split-valence basis sets arising from the group of John Pople is typically X-YZg. In this case, X represents the number of primitive Gaussians comprising each core atomic orbital basis function. The Y and Z indicate that the valence orbitals are composed of two basis functions each, the first one composed of a linear combination of Y primitive Gaussian functions, the other composed of a linear combination of Z primitive Gaussian functions. In this case, the presence of two numbers after the hyphens implies that this basis set is a split-valence double-zeta basis set. Split-valence triple- and quadruple-zeta basis sets are also used, denoted as X-YZWg, X-YZWVg, etc. Here is a list of commonly used split-valence basis sets of this type:

- 3-21G.

- 3-21G* - Polarization functions on heavy atoms.

- 3-21G** - Polarization functions on heavy atoms and hydrogen.

- 3-21+G - Diffuse functions on heavy atoms.

- 3-21++G - Diffuse functions on heavy atoms and hydrogen.

- 3-21+G* - Polarization *and* diffuse functions on heavy atoms.

- 3-21+G** - Polarization functions on heavy atoms and hydrogen, as well as diffuse functions on heavy atoms.

- 4-21G.

- 4-31G.

- 6-21G.

- 6-31G.

- 6-31G*.

- 6-31+G*.

- 6-31G(3df, 3pd).

- 6-311G.

- 6-311G*.

- 6-311+G*.

The 6-31G* basis set (defined for the atoms H through Zn) is a valence double-zeta polarized basis set that adds to the 6-31G set six *d*-type Cartesian-Gaussian polarization functions on each of the atoms Li through Ca and ten *f*-type Cartesian Gaussian polarization functions on each of the atoms Sc through Zn.

Pople basis sets are somewhat outdated, as correlation-consistent or polarization-consistent basis sets typically yield better results with similar resources. Also note that some Pople basis sets have grave deficiencies that may lead to incorrect results.

Correlation-Consistent Basis Sets

Ones of the most widely used basis sets are those developed by Dunning and coworkers, since they are designed for converging Post-Hartree–Fock calculations systematically to the complete basis set limit using empirical extrapolation techniques.

For first- and second-row atoms, the basis sets are cc-pVNZ where N=D,T,Q,5,6,... (D=double, T=triples, etc).. The 'cc-p', stands for 'correlation-consistent polarized' and the 'V' indicates they are valence-only basis sets. They include successively larger shells of polarization (correlating) functions (*d*, *f*, *g*, etc). More recently these 'correlation-consistent polarized' basis sets have become widely used and are the current state of the art for correlated or post-Hartree–Fock calculations. Examples of these are:

- cc-pVDZ - Double-zeta.

- cc-pVTZ - Triple-zeta.

- cc-pVQZ - Quadruple-zeta.

- cc-pV5Z - Quintuple-zeta, etc.

- aug-cc-pVDZ, etc. - Augmented versions of the preceding basis sets with added diffuse functions.

- cc-pCVDZ - Double-zeta with core correlation.

For period-3 atoms (Al-Ar), additional functions have turned out to be necessary; these are the cc-pV(N+d)Z basis sets. Even larger atoms may employ pseudopotential basis sets, cc-pVNZ-PP, or relativistic-contracted Douglas-Kroll basis sets, cc-pVNZ-DK.

While the usual Dunning basis sets are for valence-only calculations, the sets can be augmented with further functions that describe core electron correlation. These core-valence sets (cc-pCVXZ) can be used to approach the exact solution to the all-electron problem, and they are necessary for accurate geometric and nuclear property calculations.

Weighted core-valence sets (cc-pwCVXZ) have also been recently suggested. The weighted sets aim to capture core-valence correlation, while neglecting most of core-core correlation, in order to yield accurate geometries with smaller cost than the cc-pCVXZ sets.

Diffuse functions can also be added for describing anions and long-range interactions such as Van der Waals forces, or to perform electronic excited-state calculations, electric field property calculations. A recipe for constructing additional augmented functions exists; as many as five augmented functions have been used in second hyperpolarizability calculations in the literature. Because of the rigorous construction of these basis sets, extrapolation can be done for almost any energetic property. However, care must be taken when extrapolating energy differences as the individual energy components converge at different rates: the Hartree-Fock energy converges exponentially, whereas the correlation energy converges only polynomially.

	H-He	Li-Ne	Na-Ar
cc-pVDZ	[2s1p] → 5 func.	[3s2p1d] → 14 func.	[4s3p1d] → 18 func.
cc-pVTZ	[3s2p1d] → 14 func.	[4s3p2d1f] → 30 func.	[5s4p2d1f] → 34 func.
cc-pVQZ	[4s3p2d1f] → 30 func.	[5s4p3d2f1g] → 55 func.	[6s5p3d2f1g] → 59 func.
aug-cc-pVDZ	[Data unknown/missing.]	[Data unknown/missing.]	[Data unknown/missing.]
aug-cc-pVTZ	[Data unknown/missing.]	[Data unknown/missing.]	[Data unknown/missing.]
aug-cc-pVQZ	[Data unknown/missing.]	[Data unknown/missing.]	[Data unknown/missing.]

To understand how to get the number of functions take the cc-pVDZ basis set for H: There are two s ($L = 0$) orbitals and one p ($L = 1$) orbital that has 3 components along the z-axis (m_L = -1,0,1) corresponding to p_x, p_y and p_z. Thus, five spatial orbitals in total. Note that each orbital can hold two electrons of opposite spin.

For example, Ar [1s, 2s, 2p, 3s, 3p] has 3 s orbitals (L=0) and 2 sets of p orbitals (L=1). Using cc-pVDZ, orbitals are [1s, 2s, 2p, 3s, 3s', 3p, 3p', 3d'] (where ' represents the added in polarisation orbitals), with 4 s orbitals (4 basis functions), 3 sets of p orbitals (3 × 3 = 9 basis functions), and 1 set of d orbitals (5 basis functions). Adding up the basis functions gives a total of 18 functions for Ar with the cc-pVDZ basis-set.

Polarization-consistent Basis Sets

Density-functional theory has recently become widely used in computational chemistry. However, the correlation-consistent basis sets described above are suboptimal for density-functional theory, because the correlation-consistent sets have been designed for Post-Hartree–Fock, while density-functional theory exhibits much more rapid basis set convergence than wave function methods.

Adopting a similar methodology to the correlation-consistent series, Frank Jensen introduced polarization-consistent (pc-n) basis sets as a way to quickly converge density functional theory calculations to the complete basis set limit. Like the Dunning sets, the pc-n sets can be combined with basis set extrapolation techniques to obtain CBS values.

The pc-n sets can be augmented with diffuse functions to obtain augpc-n sets.

Karlsruhe Basis Sets

Some of the various valence adaptations of Karlsruhe basis sets are:

- def2-SV(P) - Split valence with polarization functions on heavy atoms (not hydrogen).

- def2-SVP - Split valence polarization.

- def2-SVPD - Split valence polarization with diffuse functions.

- def2-TZVP - Valence triple-zeta polarization.

- def2-TZVPD - Valence triple-zeta polarization with diffuse functions.

- def2-TZVPP - Valence triple-zeta with two sets of polarization functions.

- def2-TZVPPD - Valence triple-zeta with two sets of polarization functions and a set of diffuse functions.

- def2-QZVP - Valence quadruple-zeta polarization.

- def2-QZVPD - Valence quadruple-zeta polarization with diffuse functions.

- def2-QZVPP - Valence quadruple-zeta with two sets of polarization functions.

- def2-QZVPPD - Valence quadruple-zeta with two sets of polarization functions and a set of diffuse functions.

Completeness-optimized Basis Sets

Gaussian-type orbital basis sets are typically optimized to reproduce the lowest possible energy for the systems used to train the basis set. However, the convergence of the energy does not imply convergence of other properties, such as nuclear magnetic shieldings, the dipole moment, or the electron momentum density, which probe different aspects of the electronic wave function.

Manninen and Vaara have proposed completeness-optimized basis sets, where the exponents are obtained by maximization of the one-electron completeness profile instead of minimization of the energy. Complenetess-optimized basis sets are a way to easily approach the complete basis set limit of any property at any level of theory, and the procedure is simple to automatize.

Completeness-optimized basis sets are tailored to a specific property. This way, the flexibility of the basis set can be focused on the computational demands of the chosen property, typically yielding much faster convergence to the complete basis set limit than is achievable with energy-optimized basis sets.

Plane-wave Basis Sets

In addition to localized basis sets, plane-wave basis sets can also be used in quantum-chemical simulations. Typically, the choice of the plane wave basis set is based on a cutoff energy. The plane waves in the simulation cell that fit below the energy criterion are then included in the calculation. These basis sets are popular in calculations involving three-dimensional periodic boundary conditions.

The main advantage of a plane-wave basis is that it is guaranteed to converge in a smooth, monotonic manner to the target wavefunction. In contrast, when localized basis sets are used, monotonic convergence to the basis set limit may be difficult due to problems with over-completeness: in a large basis set, functions on different atoms start to look alike, and many eigenvalues of the overlap matrix approach zero.

In addition, certain integrals and operations are much easier to program and carry out with plane-wave basis functions than with their localized counterparts. For example, the kinetic energy operator is diagonal in the reciprocal space. Integrals over real-space operators can be efficiently carried out using fast Fourier transforms. The properties of the Fourier Transform allow a vector representing the gradient of the total energy with respect to the plane-wave coefficients to be calculated with a computational effort that scales as NPW*ln(NPW) where NPW is the number of plane-waves. When this property is combined with separable pseudopotentials of the Kleinman-Bylander type and pre-conditioned conjugate gradient solution techniques, the dynamic simulation of periodic problems containing hundreds of atoms becomes possible.

In practice, plane-wave basis sets are often used in combination with an 'effective core potential' or pseudopotential, so that the plane waves are only used to describe the valence charge density. This is because core electrons tend to be concentrated very close to the atomic nuclei, resulting in large wavefunction and density gradients near the nuclei which are not easily described by a plane-wave basis set unless a very high energy cutoff, and therefore small wavelength, is used. This combined method of a plane-wave basis set with a core pseudopotential is often abbreviated as a *PSPW* calculation.

Furthermore, as all functions in the basis are mutually orthogonal and are not associated with any particular atom, plane-wave basis sets do not exhibit basis-set superposition error. However, the plane-wave basis set is dependent on the size of the simulation cell, complicating cell size optimization.

Due to the assumption of periodic boundary conditions, plane-wave basis sets are less well suited to gas-phase calculations than localized basis sets. Large regions of vacuum need to be added on all sides of the gas-phase molecule in order to avoid interactions with the molecule and its periodic copies. However, the plane waves use a similar accuracy to describe the vacuum region as the region where the molecule is, meaning that obtaining the truly noninteracting limit may be computationally costly.

Real-space Basis Sets

Analogous to the plane wave basis sets, where the basis functions are eigenfunctions of the momentum operator, there are basis sets whose functions are eigenfunctions of the position operator, that is, points on a uniform mesh in real space. The actual implementation may use finite differences, finite elements or Lagrange sinc-functions, or wavelets.

Sinc functions form an orthonormal, analytical, and complete basis set. The convergence to the complete basis set limit is systematic and relatively simple. Similarly to plane wave basis sets, the accuracy of sinc basis sets is controlled by an energy cutoff criterion.

In the case of wavelets and finite elements, it is possible to make the mesh adaptive, so that more points are used close to the nuclei. Wavelets rely on the use of localized functions that allow for the development of linear-scaling methods.

Extended Basis Sets

There are hundreds of basis sets composed of Gaussian Type Orbitals (GTOs). The smallest of these are called minimal basis sets, and they are typically composed of the minimum number of basis functions required to represent all of the electrons on each atom. The largest of these can contain literally dozens to hundreds of basis functions on each atom.

Minimum basis Sets

A minimum basis set is one in which a single basis function is used for each orbital in a Hartree-Fock calculation on the atom. However, for atoms such as lithium, basis functions of p type are added to the basis functions corresponding to the 1s and 2s orbitals of each atom. For example, each atom in the first row of the periodic system (Li - Ne) would have a basis set of five functions (two s functions and three p functions).

In a minimum basis set, a single basis function is used for each atomic orbital on each constituent atom in the system.

The most common minimal basis set is STO-nG, where n is an integer. This nn value represents the number GTOs used to approximate the Slater Type orbital (STO) for both core and valence orbitals. Minimal basis sets typically give rough results that are insufficient for research-quality publication, but are much cheaper (less calculations requires) than the larger basis sets .Commonly used minimal basis sets of this type are: STO-3G, STO-4G, and STO-6G.

Two is often Better than One

Minimal basis sets are not flexible enough for accurate representation of, which requires the use multiple functions to represent each atomic orbital. The distribution of the electron density of valence electrons is better represented by the sum of two orbitals with different "effective charges". This is a double-ζ basis sets and includes split-valence set (inner and valence) and linear combination of two orbitals of same type, but with different effective charges (i.e., ζ). This flexibility can be used to generate atomic orbital of adjustable sizes.

For example, the double-zeta basis set allows us to treat each orbital separately when we conduct the Hartree-Fock calculation.

$$\phi i = a_1 \phi_{2s}^{STO}(r, \zeta_1) + a_2 \phi_{2s}^{STO}(r, \zeta_2)$$

The 2s atomic orbital approximated as a sum of two STOs. The two equations are the same except for the value of ζ which accounts for how large the orbital is. The constants a_1 and a_2 determines how much each STO contributes to the final atomic orbital, which will vary depending on the type of atom that the atomic orbit (i.e., hydrogen and lithium orbitals will have different a_1, a_2, ζ_1, and ζ_2 values).

The triple and quadruple-zeta basis sets work the same way, except use three and four STOs instead of two like in $\phi i = a_1 \phi_{2s}^{STO}(r, \zeta_1) + a_2 \phi_{2s}^{STO}(r, \zeta_2)$. The typical trade-off applies here as well, better accuracy, however with more expensive calculations. There are several different types of extended basis sets including: n split-valence, n polarized sets, n diffuse sets, and n correlation consistent sets.

N−MPG

For describing split-valence basis set. NN is the number of Gaussian functions describing inner-shell orbitals, while the hyphen denotes a split-valence set. M and P designate the number of Gaussian functions used to fit the two orbitals of the valence shell:

- M corresponds to number of Gaussian functions used to describe the smaller orbital.

- P corresponds to number of Gaussian functions used to describe the larger orbital (e.g., 6-31G and 3-21G).

A minimal basis set is when one basis function for each atomic orbital in the atom, while a double-ζ, has two two basis functions for each atomic orbital. Correspondingly, a triple and dquarupe-ζ set had three and four basis functions for each atomic orbital, respectively. Higher order basis set have been constructed too, e.g., 5Z, 6Z,).

There are hundreds of basis sets composed of Gaussian-type orbitals. The smallest of these are called minimal basis sets, and they are typically composed of the minimum number of basis functions required to represent all of the electrons on each atom. The largest of these can contain dozens to hundreds of basis functions on each atom.

Commonly used split-valence basis sets

3-21G	3-21G	3-21G	3-21G* - Polarized	3-21+G - Diffuse functions	3-21+G* - With polarization and diffuse functions
4-21G	4-31G	4-31G	4-31G	4-31G	
6-21G	6-31G	6-31G*	6-31+G*	6-31G(3df, 3pd)	6-311G
6-311G	6-311G*	6-311+G*			

Orbital Polarization Terms in Basis Sets

Polarization functions denoted in Pople's sets by an asterisk. Two asterisks, indicate that polarization functions are al so added to light atoms (hydrogen and helium). n Polarization functions have one additional node. For example, the only basis function located on a hydrogen at om in a minimal basis set would be a function approximating the 1 s atomic orbital. When polarization is added to this basis set, a p -function is also added to the basis set. The 6-31G** is synonymous to 6-31 G(d,p).

The use of a minimal basis set with fixed zeta parameters severely limits how much the electronic charge can be changed from the atomic charge distribution to describe molecules and chemical bonds. This limitation is removed if STOs with larger n values and different spherical harmonic functions, the $Y_l^m(\theta,\varphi)$ in the definition of STO's are included. Adding such functions is another way to expand the basis set and obtain more accurate results. Such functions are called polarization functions because they allow for charge polarization away form the atomic distribution to occur.

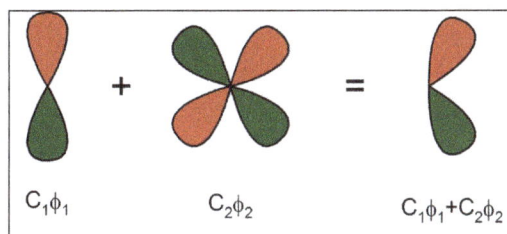

A d-polarization function added to a p orbital.

The most common addition to minimal basis sets is probably the addition of polarization functions, denoted (in the names of basis sets developed by Pople) by an asterisk, *. Two asterisks, **, indicate that polarization functions are also added to light atoms (hydrogen and helium). These are auxiliary functions with one additional node. For example, the only basis function located on a hydrogen atom in a minimal basis set would be a function approximating the 1s atomic orbital. When polarization is added to this basis set, a p-function is also added to the basis set. This adds some additional needed flexibility within the basis set, effectively allowing molecular orbitals involving the hydrogen atoms to be more asymmetric about the hydrogen nucleus.

This is an important result when considering accurate representations of bonding between atoms, because the very presence of the bonded atom makes the energetic environment of the electrons spherically asymmetric. Similarly, d-type functions can be added to a basis set with valence p orbitals, and f-functions to a basis set with d-type orbitals, and so on. Another, more precise notation indicates exactly which and how many functions are added to the basis set, such as (d, p).

Diffuse Functions

Another common addition to basis sets is the addition of diffuse functions, denoted in Pople-type sets by a plus sign, +, and in Dunning-type sets by "aug" (from "augmented"). Two plus signs indicate that diffuse functions are also added to light atoms (hydrogen and helium). These are very shallow Gaussian basis functions, which more accurately represent the "tail" portion of the atomic orbitals, which are distant from the atomic nuclei. These additional basis functions can be important when considering anions and other large, "soft" molecular systems.

References

- A-Overview-of-Quantum-Calculations, A-Computational-Quantum-Chemistry, Map%3A-Physical-Chemistry-(mcquarrie-and-Simon), Physical-and-Theoretical-Chemistry-Textbook-Maps, Bookshelves: chem.libretexts.org, Retrieved 21 August, 2019

- Errol G. Lewars (2003-01-01). Computational Chemistry: Introduction to the Theory and Applications of Molecular and Quantum Mechanics (1st ed.). Springer. ISBN 978-1402072857

- A-Orbital-Polarization-Terms-in-Basis-Sets, A-Computational-Quantum-Chemistry, Physical-and-Theoretical-Chemistry-Textbook-Maps/Map%3A-Physical-Chemistry-(mcquarrie-and-Simon), Bookshelves: chem.libretexts.org, Retrieved 22 January, 2019

- Jensen, Frank (2013). "Atomic orbital basis sets". WIREs Comput. Mol. Sci. 3 (3): 273–295. doi:10.1002/wcms.1123

- Lehtola, Susi (2015). "Automatic algorithms for completeness-optimization of Gaussian basis sets". J. Comput. Chem. 36 (5): 335–347. doi:10.1002/jcc.23802

Permissions

Index